T0216171

Lecture Notes in Artificial Intelligence   10994

Subseries of Lecture Notes in Computer Science

LNAI Series Editors

Randy Goebel
University of Alberta, Edmonton, Canada
Yuzuru Tanaka
Hokkaido University, Sapporo, Japan
Wolfgang Wahlster
DFKI and Saarland University, Saarbrücken, Germany

LNAI Founding Series Editor

Joerg Siekmann
DFKI and Saarland University, Saarbrücken, Germany

More information about this series at http://www.springer.com/series/1244

Poramate Manoonpong · Jørgen Christian Larsen
Xiaofeng Xiong · John Hallam
Jochen Triesch (Eds.)

From Animals to Animats 15

15th International Conference
on Simulation of Adaptive Behavior, SAB 2018
Frankfurt/Main, Germany, August 14–17, 2018
Proceedings

 Springer

Editors
Poramate Manoonpong (iD)
University of Southern Denmark
Odense
Denmark

Jørgen Christian Larsen (iD)
University of Southern Denmark
Odense
Denmark

Xiaofeng Xiong
University of Southern Denmark
Odense
Denmark

John Hallam
University of Southern Denmark
Odense
Denmark

Jochen Triesch
Frankfurt Institute for Advanced Studies
Frankfurt/Main
Germany

ISSN 0302-9743   ISSN 1611-3349 (electronic)
Lecture Notes in Artificial Intelligence
ISBN 978-3-319-97627-3   ISBN 978-3-319-97628-0   (eBook)
https://doi.org/10.1007/978-3-319-97628-0

Library of Congress Control Number: 2018950092

LNCS Sublibrary: SL7 – Artificial Intelligence

© Springer Nature Switzerland AG 2018
This work is subject to copyright. All rights are reserved by the Publisher, whether the whole or part of the material is concerned, specifically the rights of translation, reprinting, reuse of illustrations, recitation, broadcasting, reproduction on microfilms or in any other physical way, and transmission or information storage and retrieval, electronic adaptation, computer software, or by similar or dissimilar methodology now known or hereafter developed.
The use of general descriptive names, registered names, trademarks, service marks, etc. in this publication does not imply, even in the absence of a specific statement, that such names are exempt from the relevant protective laws and regulations and therefore free for general use.
The publisher, the authors and the editors are safe to assume that the advice and information in this book are believed to be true and accurate at the date of publication. Neither the publisher nor the authors or the editors give a warranty, express or implied, with respect to the material contained herein or for any errors or omissions that may have been made. The publisher remains neutral with regard to jurisdictional claims in published maps and institutional affiliations.

© Cover illustration: Jean Solé

This Springer imprint is published by the registered company Springer Nature Switzerland AG
The registered company address is: Gewerbestrasse 11, 6330 Cham, Switzerland

Preface

This book contains the articles presented at the 15th International Conference on the Simulation of Adaptive Behavior (SAB 2018), held at the Frankfurt Institute for Advanced Studies, Germany, in August 2018.

The objective of the biennial SAB conference is to bring together researchers in computer science, artificial intelligence, artificial life, control, robotics, neurosciences, ethology, evolutionary biology, and related fields in order to further our understanding of the behaviors and underlying mechanisms that allow natural and artificial animals to adapt and survive in uncertain environments.

Adaptive behavior research is distinguished by its focus on the modeling and creation of complete animal-like systems, which – however simple at the moment – may be one of the best routes to understanding intelligence in natural and artificial systems. The conference is part of a long series that started with the first SAB conference held in Paris in September 1990, which was followed by conferences in Honolulu (1992), Brighton (1994), Cape Cod (1996), Zürich (1998), Paris (2000), Edinburgh (2002), Los Angeles (2004), Rome (2006), Osaka (2008), Paris (2010), Odense (2012), Castellón (2014), and Aberystwyth (2016). In 1992, MIT Press introduced the quarterly journal *Adaptive Behavior*, now published by SAGE Publications. The establishment of the International Society of Adaptive Behavior (ISAB) in 1995 further underlined the emergence of adaptive behavior as a fully fledged scientific discipline. The present proceedings provide a comprehensive and up-to-date resource for the future development of this exciting field. The articles cover the main areas in animat research, including the animat approach and methodology, perception and motor control, evolution, learning, and adaptation, and collective and social behavior. The authors focus on well-defined models, computer simulations, or robotic models that help to characterize and compare various organizational principles, architectures, and adaptation processes capable of including adaptive behavior in real animals or synthetic agents, the animats.

This conference and its proceedings would not exist without the substantial help of a wide range of people. Foremost, we would like to thank the members of the Program Committee, who thoughtfully reviewed all the submissions and provided detailed suggestions on how to improve the articles. We are also indebted to our sponsors. And, once again, we warmly thank Jean Solé for the artistic conception of the SAB 2018 poster and the proceedings cover. We invite readers to enjoy and profit from the papers in this book, and look forward to the next SAB conference in 2020.

August 2018

Poramate Manoonpong
Jørgen Christian Larsen
Xiaofeng Xiong
John Hallam
Jochen Triesch

Organization

From Animals to Animats 15, the 15th International Conference on the Simulation of Adaptive Behavior (SAB 2018), was organized by Frankfurt Institute for Advanced Studies, Germany, and the International Society for Adaptive Behavior (ISAB).

General Chairs

Jochen Triesch — Frankfurt Institute for Advanced Studies, Frankfurt, Germany

John Hallam — University of Southern Denmark, Odense, Denmark

Program Chairs

Poramate Manoonpong — University of Southern Denmark, Odense, Denmark
Jørgen Christian Larsen — University of Southern Denmark, Odense, Denmark
Xiaofeng Xiong — University of Southern Denmark, Odense, Denmark

Local Organization

Eike Schädel — Frankfurt Institute for Advanced Studies, Frankfurt, Germany

Natalie Schaworonkow — Frankfurt Institute for Advanced Studies, Frankfurt, Germany

Patricia Till — Frankfurt Institute for Advanced Studies, Frankfurt, Germany

Alexander Achenbach — Frankfurt Institute for Advanced Studies, Frankfurt, Germany

Program Committee

Alexandros Giagkos — Aberystwyth University, Aberystwyth, UK
Elio Tuci — Middlesex University, London, UK
Andrew Philippides — University of Sussex, Brighton, UK
Keyan Ghazi-Zahedi — Max Planck Institute for Mathematics in the Sciences, Leipzig, Germany
Inaki Rano — Ulster University, Coleraine, UK
Elmar Rückert — Universität zu Lübeck, Lübeck, Germany
Jan-Matthias Braun — University of Southern Denmark, Odense, Denmark
Swen Gaudl — Falmouth University, Falmouth, UK
Valerio Apicella — University of Sannio, Benevento, Italy

Hugo Vieira Neto	Federal University of Technology - Paraná (UTFPR), Curitiba-PR, Brazil
Wenxiong Kang	South China University of Technology, Guangzhou, China
Neil Vaughan	Bournemouth University, Dorset, UK
Benoît Girard	Institute for Intelligent Systems and Robotics, Paris, France
Andrei D. Robu	University of Hertfordshire, Hatfield, UK
Frederic Alexandre	Neurodegeneratives Diseases Institute, Bordeaux, France
Patricia Shaw	Aberystwyth University, Aberystwyth, UK
Bernd Porr	University of Glasgow, Glasgow, UK
Cheng Hu	University of Lincoln, Lincolnshire, UK
Alastair Channon	Keele University Staffordshire, UK
Adam Stanton	Keele University Staffordshire, UK
Zhenli Lu	Changshu Institute of Technology, Changshu, China
Jean-Baptiste Mouret	Inria, CNRS, Nancy, France
Anders Christensen	University Institute of Lisbon (ISCTE-IUL), Lisbon, Portugal
Aidong Zhang	Chinese University of Hong Kong, Shenzhen, China
Tomas Kulvicius	University of Göttingen, Göttingen, Germany
Peter Eckert	École polytechnique fédérale de Lausanne, Lausanne, Switzerland
Alex Sprowitz	Max Planck Institute for Intelligent Systems, Stuttgart, Germany
Boxing Wang	Zhejiang University, Hangzhou, China
Mathias Thor	University of Southern Denmark, Odense, Denmark
Gabriel Urbain	Ghent University, Gent, Belgium
Francis Wyffels	Ghent University, Gent, Belgium
Hongkai Li	Nanjing University of Aeronautics and Astronautics, Nanjing, China
Akio Ishiguro	Tohoku University, Sendai, Japan
Hamza Khan	University of Dundee, Dundee, UK
Ryohei Nakano	Chubu University, Kasugai, Japan
Oleg Nikitin	Russian Academy of Sciences, Khabarovsk, Russia
Angelo Arleo	University Pierre&Marie Curie, Paris, France
Michael Spratling	King's College London, London, UK
Marti Sanchez-Fibla	University Pompeu Fabra, Barcelona, Spain
Frederike Kubandt	Goethe-Universität, Frankfurt am Main, Germany
Colin Johnson	University of Kent, Kent, UK
Hiroyuki Iizuka	Hokkaido University, Sapporo, Japan
Rotimi Ogunsakin	Alliance Manchester Business School, Manchester, UK
Onofrio Gigliotta	University of Naples Federico II, Napoli, Italy
Iñaki Fernández Pérez	Université de Lorraine, LORIA, Vandoeuvre-lès-Nancy Cedex, France
Muhanad Hayder	University of Kerbala, Karbala, Iraq

Jan Paul Siebert	University of Glasgow, Glasgow, UK
Silvio P. Sabatini	University of Genoa, Genova, Italy
John Nassour	Chemnitz University of Technology, Chemnitz, Germany
Marko Bjelonic	Swiss Federal Institute of Technology, Zürich, Switzerland
Julien Diard	French National Center for Scientific Research, Grenoble, France
Shinkichi Inagaki	Nagoya University, Meidai, Japan
Julien Serres	Aix-Marseille University, Marseille, France
Francois Michaud	Université de Sherbrooke, Quebec, Canada
Ruta Desai	Carnegie Mellon University, Pittsburgh, USA
Shinya Aoi	Kyoto University, Kyoto, Japan
Daniel Renjewski	Technical University of Munich, Munich, Germany
Myra S. Wilson	Aberystwyth University, Aberystwyth, UK

Sponsoring Institutions

International Society for Adaptive Behavior (ISAB)
Frankfurt Institute for Advanced Studies (FIAS)
University of Southern Denmark (SDU)
Webots (robot simulation) https://www.cyberbotics.com

Contents

Collective and Social Behavior

Animat Approach and Methodology

Impact of Mobility Mode on Innovation Dissemination: An Agent-Based Simulation Modeling

Sanad Al-Maskari, Kashif Zia[✉], Arshad Muhammad,
and Dinesh Kumar Saini

Faculty of Computing and Information Technology, Sohar University, Sohar, Oman
kzia@soharuni.edu.om

Abstract. In this paper, an agent-based model of information/innovation dissimulation and adoption, motivated by seminal model from Centola and Macy is presented. The four factors contributing to complex contagion, namely, strategic complementarity, credibility, legitimacy, and emotional contagion are considered. A parameterized mechanism of relating credibility with legitimacy in time and spatial domains is presented. Innovation dissemination in different mobility modes is studied using a proximity-based regular network. The simulation results reveal that the late adopters are affected by the early adopters, but only when the mobility model is closer to human mobility (a planned, scheduled and repeated mobility). Early adopters do not affect late adopters if all agents are stationary or acquiring random walk mobility. Also, with an increase in the percentage of early adopters, the number of people adopting an innovation increases. All the other varying factors, such as interaction radius, threshold values, etc. do not have substantial impact.

Keywords: Innovation dissemination · Agent-based model
Simulation modeling · Self-adaptive behavior

1 Introduction

Rapid diffusion of information is often associated with social media. However, information dissemination alone is not enough to make people act and bring a change. The type of information providers, a function of social networking structure, is decisive [4]. One of the important structural features in this context is the nature of ties in the network. The understanding about "the strength of weak ties" is an established fact now, given by Granovetter [7] as: "whatever is to be diffused can reach a larger number of people, and traverse a greater social distance, when passed through weak ties rather than strong". Conversely, the notion of "the weakness of strong ties" is also true resulting in localization of information diffusion, due to propagation of information only in closely knit network. It means that strong relational ties are structurally weak, and vice versa,

© Springer Nature Switzerland AG 2018
P. Manoonpong et al. (Eds.): SAB 2018, LNAI 10994, pp. 3–14, 2018.
https://doi.org/10.1007/978-3-319-97628-0_1

where relational ties are individual social ties and a structural tie represents its ability of propagating information.

Consequently, information spreads more rapidly in a small-world network structure in which a few long ties augment mostly tightly-knit local communities [18]. However, there are also some established variations; particularly evidenced in cases where there is a clear differentiation between dissemination of information and its effect on people. A thesis advocating it and in other words contradicting the notion of the "strength of weak/long ties" differentiates between mere dissemination and potentially a more demanding collective action. The seminal work is from Centola and Macy [4]. The authors postulate that "network structures that are highly efficient for the rapid dissemination of information are often not conducive to the diffusion of collective action based on the information". Authors in [4] also provide a more discrete specification capturing the soul of the argument as: "The "strength of weak ties" applies to the spread of information and disease but not to many types of social diffusion which depend on influence from prior adopters, such as participation in collective action, the use of costly innovations, or compliance with emergent norms. For these contagions, we contend that long ties are not strong in either of Granovetters meanings, relational or structural."

Information and diseases are simple contagions requiring only one source to spread. Complex contagions require two or more sources of activation. According to Centola and Macy [4], four factors contribute to a complex contagion: (i) late adopters waiting for early adopters, termed as *strategic complementarity*, (ii) *credibility* provided by neighbors who have already adopted an innovation, (iii) *legitimacy* provided by close friends who have already adopted an innovation, and (iv) "expressive and symbolic impulses in human behavior that can be communicated and amplified in spatially and socially concentrated gatherings" [5] termed as *emotional contagion*. Further, they define "a contagion as *uncontested* if activation depends solely on the number of neighbors who are activated, without regard to the number who are not activated." Whereas, a contagion is *contested* if it also depends on persons who are not activated. The implications of this, according to them, are: in case of uncontested contagions, "the larger the number of neighbors, the greater the chance of becoming activated", and in case of contested contagions, "the more neighbors an actor has, the lower the susceptibility to activation". Examples of complex contagion are spread of participation in collective action and norms and social movements. Naturally, these usually fall into the category of contested contagions.

Centola [4] extended Watts's work [18] on effectiveness of small world to note that only mere existence of a bridge between two connected components is not enough to guarantee information spreading in case of complex contagions. Rather, the bridge should have required width. But, creating long ties weakens the local propagation (within the component having source node). In their recent investigations, authors in [6] confirmed these findings in three different small-world models.

We verify this thesis using a theory-driven dissemination and acceptance model of norm propagation.

In this paper, we provide application of the Centola and Macy's information/innovation dissemination and adoption model [4] in a realistic setting. Submodels of discrete spatial configuration (a grid of cells) and of proximity-based networking are integrated with agent-based specification of the innovation adoption. Consequently, a thorough study about conditions leading to innovation dissemination and adoption is presented. Particularly, the study focus on analyzing the relationship of early and late adopters. Additionally, we quantify the relationship of late vs. early adopters in different conditions. The study also intends to quantify the relationship between credibility with legitimacy in time and spatial domains. Lastly, and most importantly, innovation dissemination in different mobility modes is studied in a proximity-based regular network.

The rest of the paper is organized as follows. In Sect. 2, a review of related work is given. In Sect. 3, a detailed description of agent-based models is given. Section 3 explains the simulation experiments and the analysis of the simulation results, followed by conclusions of the paper in Sect. 5.

2   Related Work

Social influence plays a vital role in a range of behavioral phenomena from dissemination of information to the adoption of social opinions [18]. It is often difficult to identify/analyze its exact influence due to a number of reasons. For instance people tends to engage in the same activities as their peers, thus making it difficult to identify who was influenced by whom, or even if it happened at all. Similarly, two persons may disseminate information from the same source making it difficult to identify the exact source. However, there are already some established findings. For example, people who interact more often have greater tendency to influence each other [9]. And, people who infrequently interact could have more diverse social networks, resulting in the possibility of access to novel information [3].

Social networking [8] data is considered one of the best source of information dissemination and is considered as the best platform to investigate information and innovation dissimulation. Authors in [1] used an experimental approach to quantify information sharing behavior of Facebook users. Total experimental population was 253 million of the Facebook (median and average age of subjects were 26 and 29.3). Experiment outcome showed that *weak ties* are more influential than *strong ties* in this context, in particular, the weak ties provision access to more diverse and novel information than the strong ties. In another paper, authors argued that the success of YouTube is a result of social influence [15]. Authors identified three different mechanism of social influence; these are, friends network within a community of interest, friendship ties among users from outside of a community and networks of subscribers within a community of interest. Authors highlighted that social interactions results in spread of contagion as well as increase in the magnitude of contagion.

Contrarily, with a more broader focus, authors in [13] defined "knowledge network", and emphasized that social relationships and the network relationships play a vital role in knowledge creation, diffusion and absorption. Furthermore, authors in [11] argued that strong interpersonal ties are more effective than the weak ties in order to enhance knowledge transfer and learning. Their thesis is that strong ties help to establish trust, which increase awareness to access each others knowledge. When it comes to knowledge creation, weak ties allows access to disconnected or distinct partners and results in diverse information and have a positive effect on creativity [12]. Several other examples of research on influence of tie strength on information dissemination can be listed [2, 10, 14, 16, 17].

In [16], authors proposed a distributed tie measurement mechanism by a tie strength table maintained by each mobile node which contains the social tie relationship with its correspondent nodes. Authors made use of social tie table to identify the influential and suspect members in mobile opportunistic social networks (MOSNs) to enhance the efficiency of information diffusion as number of researcher suggested that weak ties is key traits that drives the diffusion of novel social information [10, 17].

It is evidenced that weak ties tend to involve in social exchange processes of the content sharing on social websites i.e. twitter. Authors [14] found that there is 3.1% more possibility that unidirectional follower will retweet then that of bidirectional followers. [2] conducted a study on customized Facebook application that how social ties are linked to economic measure of trust using investment game with focus on the relationship among observed trust and three "revealed preference" tie strength measures [2].

These works and many others primarily focus on dissemination concept/models, possibly using real social network data, in different networking and interaction possibilities. Most of these models are analytical in nature. However, there is a lot of potential in analyzing the models using a bottom-up approach, thus, providing an opportunity to have a behavioral-based implementation at an individual level. Agent-Based Modeling (ABM) provides an approach to model a population at an individual level, with detailed spatial resolution, including the stochasticity of interactions and the mobility. Through this work, we intend to enrich relatively thin body of research done in information dissimulation modeling in ABM domain.

3   Model

The model operates in a 2D environment of 100×100 cells. The simulation runs in discrete time, each time unit termed as an iteration. At each iteration, each agent in a population of agents acts according to the model specification given below. Agents perform calculations and act in a sequence, where the order of the sequence is randomly shuffled for each iteration, thus, maintaining a fairness between agents.

Let's name the agent being in memory as **A**. Let *myneighbors* is a 2D array or table data structure (in Fig. 1, we show table columns); a list of neighbors of

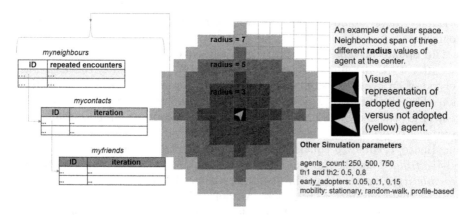

Fig. 1. Sub view of simulation space of 100×100 cells. Each agent have data structures myneighbours, mycontacts and myfriends, and its state can be adopted or not adopted. The concept of cellular radius with respect to the agent at the cell at the center of the space is also represented along with stating other simulation parameters.

A (current and before). Each neighbor is identified by its *ID* and the number of times **A** encountered it *repeated_encounters*. Since agents may be mobile, it is necessary to always seek for fresh neighborhood. To do so, we define agentset *neighbors* as all agents in prescribed **radius** excluding **A**, **radius** being one of model parameters. Then, each agent **a** in *neighbors* set is processed. If **a** already exists in table *myneighbors*, we increment *repeated_encounters* value by 1 against its ID. If **a** is encountered for the first time, we insert its ID in table *myneighbors* with *repeated_encounters* value set to 1. It is noteworthy that this process neither deletes agents from *myneighbors* if these are not in the neighborhood anymore, nor it registers the time (iteration) of an agent's encounter (first time or afterwards).

After specific number of iterations (parameter *consolidate_frequency*), we update *mycontacts* table which have a subset of agents from *myneighbors*. The table *mycontacts* have two columns (see Fig. 1). A random agent, if it already does not exist in *mycontacts*, would be added from *myneighbors* registering the iteration on which it is added against its ID. The length of *mycontacts* table have an upper bound value defined by parameter **k**. The number of agents to be added depends on length of *myneighbors* table times **k**. For example, if an agent have six neighbors, and $k = 0.5$, then the number of agents to be added in *mycontacts* table is three. It is noteworthy that as the simulation progresses, the length of *myneighbors* table increases, so does the length of *mycontacts* table. Similarly, the table *myfriends* has two columns (see Fig. 1). A random agent from *mycontacts*, if it already does not exist in *myfriends*, registering the iteration on which it is added against its ID. The length of *myfriends* table have an upper bound value defined by parameter **m**. The number of agents to be added depends on the length of *mycontacts* table times **m**. The length of *myfriends* table would also increase with *mycontacts* table as the simulation progresses.

The update of table *mycontacts* and consequently table *myfriends* happens after a specific number of iterations defined by *consolidate_frequency*. The decision of an agent to adopt or not to adopt, however, is taken at each iteration. This depends on two factors; *credibility* and *legitimacy*. *Credibility* to agent **A** is provided by neighbors who have already adopted an innovation, and *legitimacy* is provided by close friends who have already adopted an innovation. Hence, *credibility* equals %age of adopters in the current neighborhood, and *legitimacy* equals %age of adopters in table *myfriends*. More concisely, if *credibility* is greater than threshold parameter **th1** and *legitimacy* is greater than threshold parameter **th2**, agent **A** adopts the innovation.

It should be noted that we have a few early adopters in the population of agents represented by parameter **early_adopters**. In addition to mechanism of innovation adoption described above, there are three **mobility** modes under which this mechanism operates.

1. Mobility 1: No mobility; all agents are stationary.
2. Mobility 2: Random walk; agents choose a direction to move randomly at each iteration.
3. Mobility 3: Profile-based walk; agents build some random locations to move to, and they move from one location to another.

Simulation view presented in Fig. 2 depicts the difference between random-walk and profile-based mobility of an agent. Of course all agents would stationary in the third mobility mode.

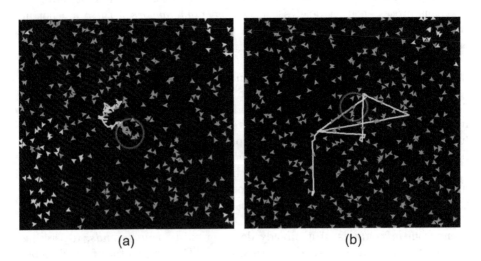

        (a)   (b)

Fig. 2. Example of (a) random-walk vs. (b) profile-based mobility of an agent

4   Simulation

4.1   Simulation Setup

NetLogo [19] is used to perform simulation experiments. The simulation space consists of 10,000 (100 × 100) cells. An agent population equal to 250 agents is randomly placed and the model of innovation dissemination is executed for many cases described by simulation parameters. Some parameters are not so relevant in the context of the study and considered as constants due to constraints of computing resources. These are consolidate_frequency (5 iterations), k (0.5) and m (0.5) and agent population (205). Whereas the varying parameters are th1, th2, agents_count, early_adopters, radius and mobility. Table 1 lists various cases generated by combinations of these parameters. Each of these cases is run for 50 times and simulation results are averaged along each of the 250 iterations of a single run.

4.2   Simulation Results

We analyze simulation results based on two parameters: (i) number of agents who have adopted (out of 250 agents), and (ii) total legitimacy value of all the agents, where each agent calculates its own legitimacy value by dividing number of agents who have adopted in the proximity divided by total number of friends. *Average Adoption* and *Average Legitimacy* both are finally achieved at each iteration by averaging 50 different simulation runs (each run having 250 iterations). The agents must attain required Credibility as well as Legitimacy to adopt (*C and L* in the graphs).

The results shown in Table 1 are summarized as:

- With increase in percentage of early adopters, the agents who have adopted (*Average Adoption*) at the end of simulation increases, irrespective of other parameters. Also, the increment from 0.05 to 0.10 to 0.15 is quite constant.
- Within one early adopters percentage, *Average Adoption* is comparable for different values of radius, sometimes with a slight increase when radius increases from 3 to 5 to 7 cells. This behavior can be related to sparse population of agents, arguing that this would not happen in denser populations, where an increase in radius would increase adoption substantially. But we saw no evidence in favor of this argument (results no presented here).
- Within one radius, there is no variation in different combinations of th1 and th2. Again, one can relate sparse population to it, arguing that this would not happen in denser populations, wherean different combinations of the values of th1 and th2 would increase/decrease adoption substantially. But we saw no evidence in favor of this argument (results no presented here).
- A comparison among three mobility models present some interesting insights. Mobility 1 and 2 (more often) produces progressive adoption (an increase in adoption happening throughout the simulation), particularly in cases from 19 to 36. It means that, in a specific setting, random walk mobility is not

Table 1. Simulation Cases with Summarized Results - 250 agents

Case	early adopters (fraction of agents)	radius (cells)	th1	th2	mobility 1	mobility 2	mobility 3	showcase
Case 1	0.05	3	0.5	0.5	12	13 †	12 ➤	✓
Case 2	0.05	3	0.5	0.8	12➤	12 †	12 ➤	
Case 3	0.05	3	0.8	0.5	12➤	13 †	12 ➤	
Case 4	0.05	3	0.8	0.8	12➤	13	12 ➤	
Case 5	0.05	5	0.5	0.5	12➤	12	12 ➤	
Case 6	0.05	5	0.5	0.8	12	12	12 ➤	
Case 7	0.05	5	0.8	0.5	12 ➤	12 ➤	12 ➤	✓
Case 8	0.05	5	0.8	0.8	12 ➤	12 ➤	12 ➤	
Case 9	0.05	7	0.5	0.5	12 ➤	12 ➤	12 ➤	
Case 10	0.05	7	0.5	0.8	12 ➤	12 ➤	12 ➤	
Case 11	0.05	7	0.8	0.5	12 ➤	12➤	12 ➤	
Case 12	0.05	7	0.8	0.8	12 ➤	12 ➤	12 ➤	
Case 13	0.10	3	0.5	0.5	25 ➤	25 ➤	25 ➤	✓
Case 14	0.10	3	0.5	0.8	25 ➤	25 ➤	25 ➤	
Case 15	0.10	3	0.8	0.5	25 ➤	25 ➤	25 ➤	
Case 16	0.10	3	0.8	0.8	25 ➤	25 ➤	25 ➤	
Case 17	0.10	5	0.5	0.5	25 ➤	25 ➤	25 ➤	
Case 18	0.10	5	0.5	0.8	25 ➤	25 ➤	25 ➤	
Case 19	0.10	5	0.8	0.5	26 ↗	29 ↗	25 ➤	✓
Case 20	0.10	5	0.8	0.8	26	28 ↗	25 ➤	
Case 21	0.10	7	0.5	0.5	26 ➤	30 ↗	25 ➤	
Case 22	0.10	7	0.5	0.8	26 ➤	28	25 ➤	
Case 23	0.10	7	0.8	0.5	26 ➤	28↗	25 ➤	
Case 24	0.10	7	0.8	0.8	26 ➤	28↗	25 ➤	
Case 25	0.15	3	0.5	0.5	37 ➤	37 ➤	37 ➤	✓
Case 26	0.15	3	0.5	0.8	37 ➤	37 ➤	37 ➤	
Case 27	0.15	3	0.8	0.5	37 ➤	37 ➤	37 ➤	
Case 28	0.15	3	0.8	0.8	37 ➤	37 ➤	37 ➤	
Case 29	0.15	5	0.5	0.5	37 ➤	37 ➤	37 ➤	
Case 30	0.15	5	0.5	0.8	37 ➤	37 ➤	37 ➤	
Case 31	0.15	5	0.8	0.5	42 ↗	46 ↗	38 ➤	✓
Case 32	0.15	5	0.8	0.8	39	43 ↗	38 ➤	
Case 33	0.15	7	0.5	0.5	41 ↗	51 ↗	39 ➤	
Case 34	0.15	7	0.5	0.8	39	43 ↗	38 ➤	
Case 35	0.15	7	0.8	0.5	39 ↗	48 ↗	39 ➤	
Case 36	0.15	7	0.8	0.8	39 ➤	40	38 ➤	

In three mobility columns, we show the final value of adoption (the number of agents who have adopted at the end of the simulation). Some of these cases are show cased in full graphs later. The values are associated with special signs, which are: progressive adoption ↗, late adoption †, early adoption ➤, and cannot decide (no sign).

much different from stationary agents. However, profile-based mobility always produces early adoption (all agents, if adopting, adopts at the early stages of the simulation). However, this is understandable and reasoned in the type of mobility that the agents have. Due to repeated inter-agent encounters, the agents build friendships faster than the other two types of mobility models, thus ensuring early adoption.

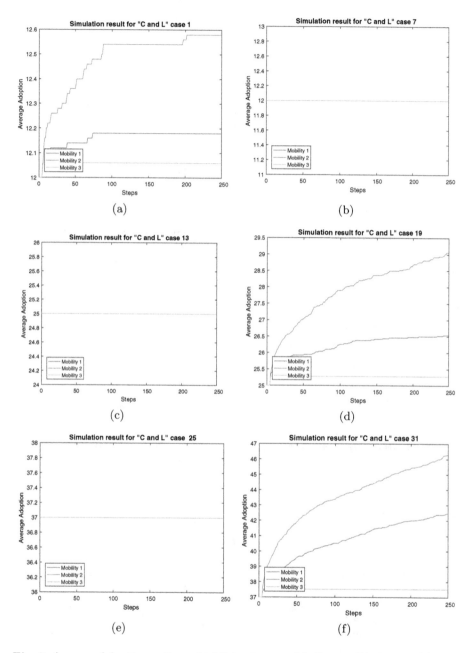

Fig. 3. Average Adoption pattern: highlighted cases. (a) Case 1; (b) Case 7; (c) Case 13; (d) Case 19; (e) Case 25; and, (f) Case 31.

Full graphs of selected cases are showcased in Fig. 3 (*Average Adoption*) and Fig. 4 (*Average Legitimacy*). A clear relationship between adoption and legitimacy is also clear from these graphs.

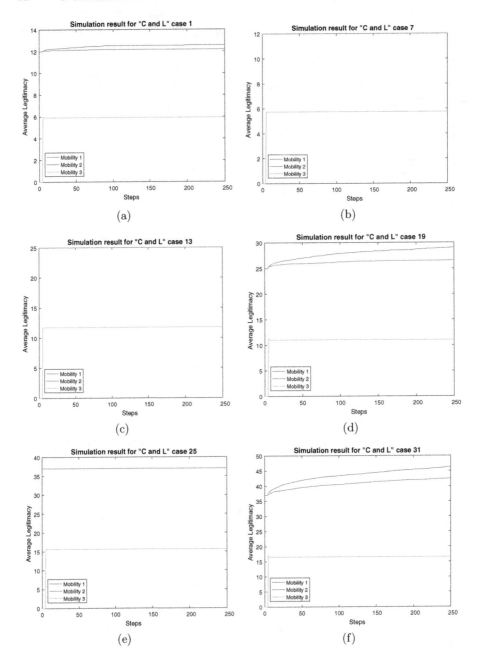

Fig. 4. Average Legitimacy pattern: highlighted cases. (a) Case 1; (b) Case 7; (c) Case 13; (d) Case 19; (e) Case 25; and, (f) Case 31.

5   Discussion and Outlook

In this paper, an agent-based model of information/innovation dissimulation and adoption, motivated from [4] is presented. The four factors contributing to complex contagion, namely, strategic complementarity, credibility, legitimacy, and emotional contagion are considered. A mechanism of relating credibility with legitimacy in time and spatial domains, tuned by parameters of interest, is presented.

The simulation results reveal that late adopters are only affected by early adopters in case of profile-based mobility. This is not applicable to stationary agents or random walk mobility. This is an important results which provides an evidence of a binding between innovation adoption and interactions. No mobility or purely random interaction does not correspond to most real-life situations. In real life, we move according to a plan and schedule, and may encounter same people again and again, thus getting affected by others. On the other hand, with an increase in the percentage of early adopters, the people adopting an innovation increases. This is expected results, evidenced in research and reality so often.

Outlook and Limitations: The research presented in this paper is part of our ongoing project about modeling of unpopular norms adherence and enforcement [21,22]. We are in process of integration of "interesting" dissemination patterns (learned in this module) with unpopular norms enforcement model of ours [20], an ideal scenario to investigate the effect of expressive and symbolic impulses onto activation dynamics. Due to this, we have following simplifications in the model: (i) We do not consider small world or any other network structure, just a regular network with neighborhood connectivity is considered. (ii) We only focus on dissemination of innovation, not on potential action. Therefore, the notion of contested vs. uncontested activation is not relevant.

References

1. Bakshy, E., Rosenn, I., Marlow, C., Adamic, L.: The role of social networks in information diffusion. In: Proceedings of the 21st international conference on World Wide Web, pp. 519–528. ACM (2012)
2. Bapna, R., Gupta, A., Rice, S., Sundararajan, A.: Trust and the strength of ties in online social networks: an exploratory field experiment. MIS Q. Manage. Inf. Syst. **41**(1), 115–130 (2017)
3. Burt, R.S.: Structural Holes: The Social Structure of Competition. Harvard University Press, Cambridge (2009)
4. Centola, D., Macy, M.: Complex contagions and the weakness of long ties. Am. J. Sociol. **113**(3), 702–734 (2007)
5. Collins, R.: Emotional energy as the common denominator of rational action. Rationality Soc. **5**(2), 203–230 (1993)
6. Ghasemiesfeh, G., Ebrahimi, R., Gao, J.: Complex contagion and the weakness of long ties in social networks: revisited. In: Proceedings of the Fourteenth ACM Conference on Electronic Commerce, pp. 507–524. ACM (2013)

7. Granovetter, M.S.: The strength of weak ties. Am. J. Sociol. **78**(6), 1360–1380 (1973)
8. Greenwood, S., Perrin, A., Duggan, M.: Social media update 2016. Pew Res. Center **11**, 83 (2016)
9. Hill, S., Provost, F., Volinsky, C.: Network-based marketing: identifying likely adopters via consumer networks. Stat. Sci. **21**(2), 256–276 (2006)
10. Jia, P., MirTabatabaei, A., Friedkin, N.E., Bullo, F.: Opinion dynamics and the evolution of social power in influence networks. SIAM Rev. **57**(3), 367–397 (2015)
11. Levin, D.Z., Cross, R.: The strength of weak ties you can trust: the mediating role of trust in effective knowledge transfer. Manage. Sci. **50**(11), 1477–1490 (2004)
12. Perry-Smith, J.E.: Social yet creative: the role of social relationships in facilitating individual creativity. Acad. Manag. J. **49**(1), 85–101 (2006)
13. Phelps, C., Heidl, R., Wadhwa, A.: Knowledge, networks, and knowledge networks: a review and research agenda. J. Manag. **38**(4), 1115–1166 (2012)
14. Shi, Z., Rui, H., Whinston, A.B.: Content sharing in a social broadcasting environment: evidence from twitter. MIS Q. **38**(1), 123–142 (2014)
15. Susarla, A., Oh, J.H., Tan, Y.: Social networks and the diffusion of user-generated content: evidence from youtube. Inf. Syst. Res. **23**(1), 23–41 (2012)
16. Wang, Y., Wu, J.: Social-tie-based information dissemination in mobile opportunistic social networks. In: 2013 IEEE 14th International Symposium and Workshops on a World of Wireless, Mobile and Multimedia Networks (WoWMoM), pp. 1–6. IEEE (2013)
17. Watts, D.J., Dodds, P.S.: Influentials, networks, and public opinion formation. J. Consum. Res. **34**(4), 441–458 (2007)
18. Watts, D.J., Strogatz, S.H.: Collective dynamics of small-worldnetworks. Nature **393**(6684), 440–442 (1998)
19. Wilensky, U.: Netlogo 4.0.4 (2008)
20. Zareen, Z., Zafar, M., Zia, K.: Conditions facilitating the aversion of unpopular norms: an agent-based simulation study. Int. J. Adv. Comput. Sci. Appl. **7**(7), 499–505 (2016)
21. Zia, K., Muhammad, A., Saini, D.K.: Agent-based simulation of socially-inspired model of resistance against unpopular norms. In: Proceedings of the 10th International Conference on Agents and Artificial Intelligence (2018)
22. Zia, K., Shaheen, M., Farooq, U., Nazir, S.: Conditions of depleting offender behavior in volunteering dilemma: an agent-based simulation study. In: Tuci, E., Giagkos, A., Wilson, M., Hallam, J. (eds.) SAB 2016. LNCS (LNAI), vol. 9825, pp. 352–363. Springer, Cham (2016). https://doi.org/10.1007/978-3-319-43488-9_31

A Probabilistic Interpretation of PID Controllers Using Active Inference

Manuel Baltieri[1,2(✉)] and Christopher L. Buckley[1,2]

[1] Evolutionary and Adaptive Systems Group, Department of Informatics,
University of Sussex, Brighton, UK
m.baltieri@sussex.ac.uk
[2] Sussex Neuroscience, University of Sussex, Brighton, UK

Abstract. In the past few decades, probabilistic interpretations of brain functions have become widespread in cognitive science and neuroscience. The Bayesian brain hypothesis, predictive coding, the free energy principle and active inference are increasingly popular theories of cognitive functions that claim to unify understandings of life and cognition within general mathematical frameworks derived from information and control theory, statistical physics and machine learning. The connections between information and control theory have been discussed since the 1950's by scientists like Shannon and Kalman and have recently risen to prominence in modern stochastic optimal control theory. However, the implications of the confluence of these two theoretical frameworks for the biological sciences have been slow to emerge. Here we argue that if the active inference proposal is to be taken as a general process theory for biological systems, we need to consider how existing control theoretical approaches to biological systems relate to it. In this work we will focus on PID (Proportional-Integral-Derivative) controllers, one of the most common types of regulators employed in engineering and more recently used to explain behaviour in biological systems, e.g. chemotaxis in bacteria and amoebae or robust adaptation in biochemical networks. Using active inference, we derive a probabilistic interpretation of PID controllers, showing how they can fit a more general theory of life and cognition under the principle of (variational) free energy minimisation under simple linear generative models.

1   Introduction

Probabilistic approaches to the study of living systems and cognition are becoming increasingly popular in the natural sciences. In particular for the brain sciences, theories inspired the Bayesian brain hypothesis such as predictive coding, the free energy principle and active inference have been used to explain brain processes including perception, action and higher order functions [9,12,15,22,28,31,33,34].

© Springer Nature Switzerland AG 2018
P. Manoonpong et al. (Eds.): SAB 2018, LNAI 10994, pp. 15–26, 2018.
https://doi.org/10.1007/978-3-319-97628-0_2

According to these theories, brains, and biological systems in general, should be thought of as Bayesian inference machines, gathering and representing information from the environment into generative models [12,22,28]. Such systems in fact appear to estimate the latent causes of their sensory input in a process consistent with a Bayesian inference scheme. In particular it has been suggested that perceptual process can be accounted for in terms of predictive coding models whereby feedforward prediction errors and feedback predictions are combined under a generative model to infer the hidden causes of sensory data [33]. More recent theories have extended this proposal to account also for motor control and behaviour [19,26]. On this view, behaviour is cast as a process of acting on the world to make sensory data better fit existing predictions. This latter process usually falls under the name of *active inference*.

Modelling approaches inspired by control theory are nowadays established methodologies for instance in psychology [10,32], and are increasingly popular in fields such as biology [1,41,42]. Typically inspired by classical control theory and dynamical system theory, they emphasise regulation and concepts such as set-point control and negative feedback for the study of different aspects of living systems, an approach first introduced with cybernetics [4,39]. In particular, methods such as PID (Proportional-Integral-Derivative) control have been widely used as they represent a very simple methodology with properties that guarantee robustness to perturbations and noise [1,35,41,42].

The relationship between information/probability theory and control theory has long been recognised, with the first intuitions (as far as the authors are aware) emerging from work by Shannon [37] and Kalman [30]. A unifying view of these two theoretical frameworks is nowadays proposed for instance in stochastic optimal control [38] and active inference [20,36]. How these ideas can be used to combine traditional concepts of control more commonly applied in biology with frameworks like active inference is still however unclear. Here, to address this, we develop an information theoretic interpretation of PID controllers, a very popular control strategy that works with little prior knowledge of the process to regulate. Starting from ideas proposed by the free energy principle, we will show that simple linear generative models approximating the true dynamics of the environment implicitly implement PID control as a process of active inference.

2   The Free Energy Principle

The free energy principle (FEP) was initially introduced by Friston [22] and later elaborated in a series of papers [18,19,24]. The FEP is proposed as a unifying theory for biological sciences with roots in information theory, thermodynamics and statistical mechanics. Work on the FEP has so far covered computational neuroscience [16,17], and behavioural/cognitive neuroscience studies [23,26]. Furthermore, connections have been implied with theories of biological self-organisation, information theory (e.g. infomax principle), optimal control, cybernetics and economics (e.g. utility theory) among others [19–21,36]. According to the FEP, a living system exists only in a limited set of states over time, e.g. a fish can't survive for long out of water. Biological creatures can thus be seen as systems that minimise the surprisal (or surprise/self-information) of their sensory observations to maintain their existence, e.g. a fish' observations should be limited to states in the water. Since this surprisal is not directly accessible by an agent [22], variational free energy is proposed as a proxy that can be minimised in its place, acting as an upper bound for such quantity.

In this study we focus on hypotheses and theories linked to the FEP regarding perception and action in agents, in particular predictive coding and active inference. Predictive coding [33] models of information processing in the brain prescribe a way in which top-down and bottom-up information flows could be combined in the cortex under deep generative models and are, on this view, often described a special case of the FEP [19]. Top-down processes provide predictions about sensory input while bottom-up activity carries prediction errors representing the difference between real and predicted sensations. These errors are then used to train a generative model to produce better predictions. The minimisation of prediction errors achieved by updating the predictions of this model to better represent an agent's sensations corresponds, on this view, to perception. However, one of the main contributions of the FEP is the extension of predictive coding models to include an account of action, known as *active inference* [25, 26], and thus unify perception and action in a single cohesive mathematical framework where differences between the two processes almost vanish.

2.1   Active Inference and Control

Active inference provides a second way in which prediction errors, or free energy, can be minimised. While perceptual inference suppresses prediction errors only by updating predictions of a generative model of the incoming sensations [18, 26], active inference minimises errors also by directly acting on the environment to change sensory input to better accord with existing predictions. If a generative model encodes information about favourable states for an agent, then this process constitutes a way by which an agent can change its environment to better meet its needs. Thus, under the FEP, these two processes of error suppression allow an agent to both perceive and control the surrounding environment.

Most agent-based models implementing the FEP and active inference assume that agents have a deep understanding of their environment and its dynamics in the form of an accurate and detailed generative model. For instance, in [25, 26] the generative model of the agent explicitly mirrors the *generative process* of the environment, i.e. the dynamics of the world the agent interacts with. In recent work we have argued that this needs not be the case [7], especially if we consider simple living systems with limited resources. We intuitively don't expect an ant to model the entire environment where it forages, performing complex simulations of the world in its brain (cf. the concept of Umwelt [11]). This idea is however common in other work [23, 25, 26, 28], where cognition and perception are presented as processes of inference to the best explanation, encoding an accurate set of parameters and variables of the environment with agents seen as mainly building sophisticated models of their worlds then used for action and behaviour.

One often implicit assumption is that *all* the information available to an agent should be encoded (e.g. an ant modelling the entire environment). This is however not reasonable in dynamic and complex environments: when variables and parameters in the world change too rapidly, accurate online inference and learning are implausible [3]. A possible alternative introduces action-oriented models entailing a more parsimonious approach where only task-relevant information is encoded [13,14]. On this view, agents only model environmental properties that are necessary for their behaviour and their goals.

In this work we present an example of such parsimonious, action-oriented model described in [13,14], connecting them to methods from classic control theory. We focus in particular on Proportional-Integral-Derivative (PID) control, both extensively used in industry [6] and more recently emerging as a model of robust feedback mechanisms in biology, implemented for instance by bacteria [42], amoeba [41] and gene networks [1], and in psychology [35]. PID controllers are ubiquitous in engineering mostly due to the fact that one needs only little knowledge of the process to regulate. In active inference terms, we will show that this corresponds to linear generative models that only approximate properties of the world dynamics. Specifically, our model will describe linear dynamics for a single hidden state and a linear mapping from the hidden state to an observed variable, representing knowledge of the world that is potentially far removed from the real complexity behind observations and their hidden causes.

3   PID Control

Proportional-Integral-Derivative (PID) control is one of the simplest set-point controllers, whereby a desired state (i.e. set-point, reference) represents the final goal of the regulation process, e.g. to maintain a room temperature of 23° C. PID controllers are based on closed-loop strategies with a negative feedback mechanism that tracks the real state of the environment. The difference between such state and the target value (e.g. 23° C temperature) produces a prediction error whose minimisation drives the controller, e.g. if the temperature is too high, it is decreased and if too low, it is raised. In mathematical terms:

$$e(t) = y_r - y(t) \tag{1}$$

where $e(t)$ is the error, y_r is the *reference* or set-point (e.g. desired temperature) and $y(t)$ is the observed variable (e.g. the actual room temperature).

This mechanism is however unstable in very common conditions, in particular when a steady-state offset is added (e.g. a sudden and unpredictable change in external conditions affecting the room temperature which are not under our control), or when fluctuations need to be repressed (e.g. too many oscillations in the temperature on the trajectory to the desired state may be undesirable). PID controllers deal with both of these problems by augmenting the standard negative feedback architecture, here called *proportional* or *P term*, with an *integral* or *I* and a *derivative* or *D term*, see Fig. 1. The integral term accumulates the prediction error over time in order to cancel out errors due to steady-state input,

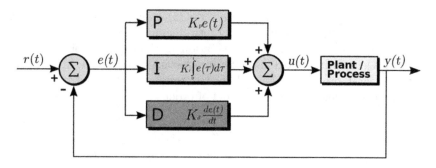

Fig. 1. A PID controller [2]. The prediction error $e(t)$ is given by the difference between a reference signal $r(t)$, y_r in our formulation, and the output $y(t)$ of a process. The different terms, one proportional to the error (P term), one integrating the error over time (I term) and one differentiating it (D term), drive the control signal $u(t)$.

while minimising the derivative of the prediction error leads to a decrease in the amplitude of fluctuations. The general form of the control signal $u(t)$ generated by a PID controller is usually described by:

$$u(t) = k_p e(t) + k_i \int_0^t e(\tau)d\tau + k_d \frac{de(t)}{dt} \tag{2}$$

where $e(t)$ is again the prediction error and k_p, k_i, k_d are the so called proportional, integral and derivative gains respectively, a set of parameters used to tune the relative strength of the P, I and D terms of the controller.

The popularity of PID controllers is largely due to: (1) their robustness in the presence of uncertainty, i.e. step disturbances and more in general noise, given by the filtering properties of the I term, and (2) an only approximate model of the dynamics of the process to regulate, based on a linearisation around the target state. This might look incompatible with standard work in active inference formulations suggesting a link to optimal control strategies with perfect models of process/environment, however, we argued previously that this needs not be the case [7]. Indeed one of the main strengths of active inference lies, according to us, in its general formulation and in generative models that do not have to mirror the dynamics of the entire environment.

4   PID Control as Active Inference

In this work we will not provide a complete derivation of the active inference scheme, referring to previous treatments [9, 17, 27] for more details. We will begin from the Laplace encoded variational free energy for a univariate case:

$$F = -\ln P(\rho, x) + constants \tag{3}$$

where ρ represents the observed sensory input of an agent and x encodes the expectation of hidden states in the environment. The remaining constants will not be discussed since they play no role in the minimisation scheme we present.

As previously shown [9,17], to minimise Eq. (3) we need to specify the agent's generative density $P(\rho, x) = P(\rho|x)P(x)$ introducing a likelihood $P(\rho|x)$ and a prior $P(x)$ in terms of an agent's expectations x. These probabilities can be specified by a generative model in the form of a state space model. In particular, to get the integral, proportional and derivative terms of a PID controller, we will use a *generalised* linear state space model of order 2 [9,17]:

$$
\begin{aligned}
\rho &= x + z & \dot{x} &= x' = -\alpha(x + \eta + w) \\
\rho' &= x' + z' & \dot{x}' &= x'' = -\alpha(x' + \eta' + w') \\
\rho'' &= x'' + z'' & \dot{x}'' &= x''' = -\alpha(x'' + \eta'' + w'')
\end{aligned}
\tag{4}
$$

where ρ is the observation of the proprioceptive signal and x is the estimated hidden state, α encodes the decay of x (α is here assumed to be very large, meaning that the generative model represents a belief in an environment that quickly relaxes to equilibrium), η encodes a desired state (e.g. desired temperature) represented mathematically as an exogenous input (or as a prior from higher layers in hierarchical models [17]) and z, w are zero-mean Gaussian random variables. As we shall see later, η is equivalent to y_r in Eq. (1). However, unlike standard PID schemes, η is here specified as a function of time using generalised coordinates of motion (explained below). Ultimately, this derivation will collapse to a more standard set-point scheme when $\eta = y_r$ and $\eta' = \eta'' = 0$. The prime (e.g. ρ', ρ'') indicates the order of generalised coordinate of motion [9,17], which are introduced to represent non-Markovian continuous stochastic processes [29], in our case z, w. One could think of them as quantities conveying information about "velocity" (e.g. ρ'), "acceleration" (e.g. ρ''), etc. for each variable. Following [9,17], we then define:

$$
\tilde{\rho} = \tilde{x} + \tilde{z} \qquad \dot{\tilde{x}} = -\alpha(\tilde{x} + \tilde{\eta} + \tilde{w})
\tag{5}
$$

where the tilde sign (e.g. $\tilde{\rho}$) summarises the generalised state, a variable and its higher orders of motion, into a more compact description (e.g. $\tilde{\rho} = \{\rho, \rho', \rho'', \dots\}$).

With the assumption that random variables $\tilde{z}, \tilde{w}$ are normally distributed (making the likelihood $P(\tilde{\rho}|\tilde{x})$ and the prior $P(\tilde{x})$ of Gaussian form), the variational free energy reduces to:

$$
\begin{aligned}
F = \frac{1}{2}\Big[&\pi_z\big(\rho - \mu_x\big)^2 + \pi_{z'}\big(\rho' - \mu_x'\big)^2 + \pi_{z''}\big(\rho'' - \mu_x''\big)^2 + \pi_w\big(\mu_x' + \alpha(\mu_x - \eta)\big)^2 \\
&+ \pi_{w'}\big(\mu_x'' + \alpha(\mu_x' - \eta')\big)^2 + \pi_{w''}\big(\mu_x''' + \alpha(\mu_x'' - \eta'')\big)^2 - \ln\big(\pi_z\pi_w\pi_{z'}\pi_{w'}\pi_{z''}\pi_{w''}\big) \Big]
\end{aligned}
\tag{6}
$$

where we used the means $\tilde{\mu}_x$ of the estimated hidden states rather than the states $\tilde{x}$ themselves since $\tilde{\mu}_x$ are the only sufficient statistics required for the minimisation of free energy under the assumption of optimal (co)variances of the recognition density (see [9,17]). $\pi_{\tilde{z}}, \pi_{\tilde{w}}$ are the precision parameters (inverse variances) of $\tilde{z}, \tilde{w}$ respectively. Following [17,27], the optimisation of the Laplace encoded free energy can be performed via a standard gradient descent procedure:

$$
\dot{\tilde{\mu}}_x = D\tilde{\mu}_x - \frac{\partial F}{\partial \tilde{\mu}_x} \qquad \dot{a} = -\frac{\partial F}{\partial a} = -\frac{\partial F}{\partial \tilde{\rho}}\frac{\partial \tilde{\rho}}{\partial a}
\tag{7}
$$

where the two equations prescribe perception and action processes respectively. The first equation includes an extra term $D\tilde{\mu}_x$ that represents the "mean of the motion" in the minimisation of variables in generalised coordinates of motion [9,17], with D as a differential operator with respect to time, i.e. $D\tilde{\mu}_x = \tilde{\mu}'_x$. More intuitively, since we are now minimising the components of a generalised state representing a trajectory rather than a static variable, variables are in a moving framework of reference where the minimisation is achieved for $\dot{\tilde{\mu}}_x = \tilde{\mu}'_x$ rather than $\dot{\tilde{\mu}}_x = 0$. At this point, the mean of the motion becomes the motion of the mean, thereby satisfying Hamilton's principle of least action [17]. In the second equation, an assumption of active inference is that actions a only affect sensory input $\tilde{\rho}$ and furthermore that this mapping is known to the agent and enacted as a reflex mechanism, see [26] for discussion. By applying the gradient descent described in Eq. (7) to our free energy function in Eq. (6), we then obtain the following update equations for perception:

$$\dot{\mu}_x = \mu'_x - \left[-\pi_z \left(\rho - \mu_x \right) + \pi_w \alpha \left(\mu'_x + \alpha(\mu_x - \eta) \right) \right]$$

$$\dot{\mu}'_x = \mu''_x - \left[-\pi_{z'} \left(\rho' - \mu'_x \right) + \pi_{w'} \alpha \left(\mu''_x + \alpha(\mu'_x - \eta') \right) + \pi_w \left(\mu'_x + \alpha(\mu_x - \eta) \right) \right] \qquad (8)$$

$$\dot{\mu}''_x = \mu'''_x - \left[-\pi_{z''} \left(\rho'' - \mu''_x \right) + \pi_{w''} \alpha \left(\mu'''_x + \alpha(\mu''_x - \eta'') \right) + \pi_{w'} \left(\mu''_x + \alpha(\mu'_x - \eta') \right) \right]$$

and for action:

$$\dot{a} = -\left[\pi_z \left(\rho - \mu_x \right) \frac{\partial \rho}{\partial a} + \pi_{z'} \left(\rho' - \mu'_x \right) \frac{\partial \rho'}{\partial a} + \pi_{z''} \left(\rho'' - \mu''_x \right) \frac{\partial \rho''}{\partial a} \right] \qquad (9)$$

The mapping of these equations to a PID control scheme becomes more clear under a few simplifying assumptions, starting from an agent that will have strong priors (desires) on the causes of its proprioceptive observations. Intuitively, these priors will be used to define actions that change the observations to better fit the agent's desires. This is implemented in the weighting mechanism of prediction errors by precision parameters in Eq. (6); see also [7,8,26] for similar discussions on precisions and behaviour. Here we want to weight prediction errors on the expected causes, $\pi_{\tilde{w}}(\tilde{\mu}'_x + \tilde{\mu}_x - \tilde{\eta})$, more than the ones on observations, $\pi_{\tilde{z}}(\tilde{\rho} - \tilde{\mu}_x)$. To achieve this, we decrease precisions $\pi_{\tilde{z}}$ on proprioceptive observations, effectively biasing the gradient descent procedure towards minimising errors on the priors [8]. We then set the decay parameter α to a large value, meaning that the agent encodes beliefs in a world that quickly settles to an equilibrium state, with higher orders of generalised motion in each line of Eq. (8) not considered during the minimisation. Perception is then approximated as:

$$\dot{\mu}_x \approx \mu'_x - \pi_w \alpha \left(\mu'_x + \alpha(\mu_x - \eta) \right)$$

$$\dot{\mu}'_x \approx \mu''_x - \pi_{w'} \alpha \left(\mu''_x + \alpha(\mu'_x - \eta') \right) \qquad (10)$$

$$\dot{\mu}''_x \approx \mu'''_x - \pi_{w''} \alpha \left(\mu'''_x + \alpha(\mu''_x - \eta'') \right)$$

where $\mu_x''' = 0$ since we truncated our generalised state-space model to order 2 (i.e. anything beyond that is zero-mean Gaussian noise). This system of equations sets, at steady state, the expected hidden states $\tilde{\mu}_x$ to the priors $\tilde{\eta}$, $\tilde{\mu}_x = \tilde{\eta}$.

To minimise free energy in presence of strong priors, the agent will necessarily have to modify its sensory input $\tilde{\rho}$ to better match expectations $\tilde{\mu}_x$ which in turn will be shaped by the priors (i.e. desires) $\tilde{\eta}$. Effectively, the agent "imposes" its desires on the world, driving actions that will minimise the prediction errors arising at the proprioceptive sensory layers. In essence, an active inference agent implements set-point regulation by acting to make its sensations accord with its strong priors/desires. After these assumptions, action can be written as:

$$\dot{a} \approx - \left[\pi_z \left(\rho - \eta \right) \frac{\partial \rho}{\partial a} + \pi_{z'} \left(\rho' - \eta' \right) \frac{\partial \rho'}{\partial a} + \pi_{z''} \left(\rho'' - \eta'' \right) \frac{\partial \rho''}{\partial a} \right] \qquad (11)$$

where we assumed $\tilde{\mu}_x = \tilde{\eta}$, but still need to specify partial derivatives $\partial \tilde{\rho}/\partial a$. As discussed previously [26], this step also highlights the fundamental differences between the FEP and the more traditional forward/inverse models formulation of control problems [40]. While these derivatives define a form of inverse model, unlike more traditional approaches this does not involve a mapping between actions and hidden states $\tilde{x}$ but is cast in terms of sensory data $\tilde{\rho}$ directly. It is claimed that this provides an easier implementation for such an inverse model [20], one that is grounded in an extrinsic frame of reference (i.e. the real world, $\tilde{\rho}$) rather than in a intrinsic one in terms of hidden states $\tilde{x}$. To achieve PID-like control, we finally assume that the agent adopts the simplest (i.e. linear) relationship between its actions (controls) and their effects on sensory input across all generalised coordinates of motion:

$$\frac{\partial \rho}{\partial a} = \frac{\partial \rho'}{\partial a} = \frac{\partial \rho''}{\partial a} = 1 \qquad (12)$$

This reflects a very simple reflex-arc-like mechanism that is triggered any time a proprioceptive prediction is made. Intuitively, positive actions increase the values of the sensed variables $\tilde{\rho}$, while negative actions decrease them. There is however an apparent inconsistency here that we need to dissolve: the proprioceptive input ρ and its higher order states ρ', ρ'' are *all* linearly dependent on actions a as represented in Eq. (12). While an action may not change position, velocity and acceleration of a variable in the same way, the goal of an agent is not to perfectly represent its physical reality but just to encode sensorimotor properties that allow it to achieve its goals. In the same way, PID controllers are, in most cases, effective but only approximate solutions for control [5]. This allows us to understand the encoding of an inverse model from the perspective of an agent rather than assuming a perfect, objective mapping from sensations to actions that reflects exactly how actions affect sensory input [26]. This also points at possible investigations of generative/inverse models in simpler living systems where accurate internal models are not needed, and where strategies like PID control are implemented [1,41,42]. By combining Eqs. (11) and (12), action can then be simplified to:

$$\dot{a} \approx \pi_z \left(\eta - \rho \right) + \pi_{z'} \left(\eta' - \rho' \right) + \pi_{z''} \left(\eta'' - \rho'' \right) \qquad (13)$$

consistent with the "velocity form" or algorithm of a PID controller [5]:

$$\dot{u} = k_i(y_r - y) + k_p\frac{d}{dt}(y_r - y) + k_d\frac{\dot{d}^2}{dt^2}(y_r - y) \tag{14}$$

where we removed the explicit dependence on time t. Velocity forms are used in control problems where, for instance, integration is provided by an external mechanism outside the controller [5,6]. This algorithm is often described using discrete systems to avoid the definition of the derivative of random variables, often assumed to be white noise (i.e. Markov processes). In the continuous case, if the variable y is a Markov process, its time derivative is in fact not well defined. For this form to exist in continuous systems, y must be a smooth process. This effectively drops the Markov assumption of white noise and implements the same definition of analytic (i.e. differentiable) noise and related generalised coordinates of motion we described earlier. The presence of extra prediction errors beyond the traditional negative feedback (proportional term) can in this light be seen as a natural consequence of considering non-Markov processes. To ensure that the active inference implementation approximates the velocity form of PID control we then need to clarify the relationship between generalised coordinates of motion in Eq. (13) and the differential operators $d/dt, d^2/dt^2$ in Eq. (14). As pointed out in previous work [9,27], the two of them are equal at the minimum of the free energy landscape, when the gradient descent has reached its steady state. To simplify our formulation and show this more directly, we could consider the case for $\eta' = \eta'' = 0$, defining the more standard set-point control where the desired trajectory collapses to a single point, equivalent in the velocity form to the case where y_r is a constant and $dy_r/dt = d^2y_r/dt^2 = 0$.

5   Conclusion

PID controllers are robust controllers used as a model of regulation for noisy and non-stationary processes in different disciplines, from engineering to biology and psychology. They however do not guarantee optimality, so a straightforward interpretation of this control strategy in terms of optimal control is missing. Active inference is often described as an extension of optimal control theory with deep connections to Bayesian inference [20]. While active inference has been proposed as a general mathematical theory of life and cognition according to the minimisation of variational free energy [19], methods such as PID control are still widely adopted as models of biological systems [1,41,42]. In this work we proposed a way to connect these two perspectives showing how PID controllers can be seen as a special case of active inference once simplified (i.e. linear) generative models are introduced. The ubiquitous efficacy of PID control may thus reflect the fact that the simplest models of controlled dynamics are first-order approximations to generalised motion. This simplicity is mandated because the minimisation of free energy is equivalent to the maximisation of model evidence, which can be expressed as accuracy minus complexity [14,19]. On this view, PID control emerges via the implementation of parsimonious (minimum complexity) generative models that are the most effective (maximum accuracy) for a task.

Following our previous work [7], we defined a generative model that only approximates the agent's environment and showed how under a set of assumptions including analytic (i.e. non-Markovian, differentiable) Gaussian noise and linear dynamics, this model recapitulates PID control. A crucial component of our formulation is the presence of low precision parameters on proprioceptive prediction errors of our free energy function or equivalently, beliefs about high variance of proprioceptive signals. These low precisions play two roles during the minimisation of free energy: (1) they implement control signals as predictions of proprioceptive input influenced by strong priors (i.e. desires) rather than by observations, see Eq. (11) and [7,26], and (2) they reflect a belief into the presence of large exogenous fluctuations (low precision = high variance) as part of the observed proprioceptive input. This last point can be seen as the well known property of the Integral term [6] of PID controllers, dealing with unexpected external input (i.e. large exogenous fluctuations). The model represented by derivatives $\partial\tilde{\rho}/\partial a$ encodes then how actions a approximately affect observed proprioceptive sensations $\tilde{\rho}$, with an agent implementing a sensorimotor mapping that does not match the real dynamics of actions applied to the environment. The generative model we proposed can in general be applied to different tasks, in the same way PID control is used in different problems without specific knowledge of the system to regulate.

In future work we will explore the implications of PID control as active inference for the study of biological systems. In particular we suggest that given our formalisation it is trivial to generalise the set-point definition of PID controllers, based on point attractors, to trajectories (e.g. a reference temperature changing during the day with pre-specified properties such as the rate of change, etc.) using generalised coordinates of motion [9,27]. We may also be able to provide a Bayes-optimal algorithm for the optimisation of the gains of a PID controller, k_i, k_p, k_d (i.e. precision parameters in our free energy formulation, $\pi_w, \pi_{w'}, \pi_{w''}$), for which only heuristic methods exist at the moment [5].

Acknowledgements. The authors would like to thank Karl Friston for thought-provoking discussions and insightful feedback on the final version of this manuscript, and Martijn Wisse and Sherin Grimbergen for important comments on the mathematical derivation.

References

1. Ang, J., Bagh, S., Ingalls, B.P., McMillen, D.R.: Considerations for using integral feedback control to construct a perfectly adapting synthetic gene network. J. Theor. Biol. **266**(4), 723–738 (2010)
2. Urquizo, A.: PID controller - Wikipedia, the free encyclopedia (2011). Accessed 30 Mar 2018
3. Ashby, W.: Requisite variety and its implications for the control of complex systems. Cybernetica **1**, 83–99 (1958)

4. Ashby, W.R.: An Introduction to Cybernetics. Chapman & Hall Ltd., London (1957)
5. Åström, K.J.: PID Controllers: Theory, Design and Tuning. Instrument society of America (1995)
6. Åström, K.J., Murray, R.M.: Feedback Systems: An Introduction for Scientists and Engineers. Princeton University Press, Princeton (2010)
7. Baltieri, M., Buckley, C.L.: An active inference implementation of phototaxis. In: Proceedings of European Conference on Artificial Life, pp. 36–43 (2017)
8. Brown, H., Adams, R.A., Parees, I., Edwards, M., Friston, K.: Active inference, sensory attenuation and illusions. Cogn. Process. **14**(4), 411–427 (2013)
9. Buckley, C.L., Kim, C.S., McGregor, S., Seth, A.K.: The free energy principle for action and perception: a mathematical review. J. Math. Psychol. **14**, 55–79 (2017)
10. Carver, C.S., Scheier, M.F.: Attention and self-regulation: a control-theory approach to human behavior. Springer, New York (1981)
11. Clark, A.: Being There: Putting Brain, Body, and World Together Again. MIT press, Cambridge (1998)
12. Clark, A.: Whatever next? Predictive brains, situated agents, and the future of cognitive science. Behav. Brain Sci. **36**(03), 181–204 (2013)
13. Clark, A.: Radical predictive processing. South. J. Philos. **53**(S1), 3–27 (2015)
14. Clark, A.: Surfing Uncertainty: Prediction, Action, and the Embodied Mind. Oxford University Press, New York (2015)
15. Dayan, P., Hinton, G.E., Neal, R.M., Zemel, R.S.: The Helmholtz machine. Neural Comput. **7**(5), 889–904 (1995)
16. Friston, K.: A theory of cortical responses. Philos. Trans. R. Soc. London Ser. B Biol. Sci. **360**(1456), 815–836 (2005)
17. Friston, K.: Hierarchical models in the brain. PLoS Comput. Biol. **4**(11), e1000211 (2008)
18. Friston, K.: The free-energy principle: a rough guide to the brain? Trends Cogn. Sci. **13**(7), 293–301 (2009)
19. Friston, K.: The free-energy principle: a unified brain theory? Nat. Rev. Neurosci. **11**(2), 127–138 (2010)
20. Friston, K.: What is optimal about motor control? Neuron **72**(3), 488–498 (2011)
21. Friston, K.: Life as we know it. J. R. Soc. Interface **10**(86), 20130475 (2013)
22. Friston, K., Kilner, J., Harrison, L.: A free energy principle for the brain. J. Physiol. Paris **100**(1), 70–87 (2006)
23. Friston, K., Mattout, J., Kilner, J.: Action understanding and active inference. Biol. Cybern. **104**(1–2), 137–160 (2011)
24. Friston, K., Rigoli, F., Ognibene, D., Mathys, C., Fitzgerald, T., Pezzulo, G.: Active inference and epistemic value. Cogn. Neurosci. **6**(4), 1–28 (2015)
25. Friston, K.J., Daunizeau, J., Kiebel, S.J.: Reinforcement learning or active inference? PLoS One **4**(7), e6421 (2009)
26. Friston, K.J., Daunizeau, J., Kilner, J., Kiebel, S.J.: Action and behavior: a free-energy formulation. Biol. Cybern. **102**(3), 227–260 (2010)
27. Friston, K.J., Trujillo-Barreto, N., Daunizeau, J.: DEM: a variational treatment of dynamic systems. NeuroImage **41**(3), 849–885 (2008)
28. Hohwy, J.: The Predictive Mind. OUP, Oxford (2013)
29. Jazwinski, A.H.: Stochastic Processes and Filtering Theory, vol. 64. Academic Press, New York (1970)
30. Kalman, R.E., et al.: Contributions to the theory of optimal control. Bol. Soc. Mat. Mexicana **5**(2), 102–119 (1960)

31. Knill, D.C., Pouget, A.: The Bayesian brain: the role of uncertainty in neural coding and computation. Trends Neurosci. **27**(12), 712–719 (2004)
32. Powers, W.T.: Behavior: The Control of Perception. Aldine, Chicago (1973)
33. Rao, R.P., Ballard, D.H.: Predictive coding in the visual cortex: a functional interpretation of some extra-classical receptive-field effects. Nature Neurosci. **2**(1), 79–87 (1999)
34. Rafal, B.: A tutorial on the free-energy framework for modelling perception and learning. J. Math. Psychol. **76**, 198–211 (2017)
35. Ritz, H., Nassar, M.R., Frank, M.J., Shenhav, A.: A control theoretic model of adaptive behavior in dynamic environments. bioRxiv, p. 204271 (2017)
36. Seth, A.K.: The cybernetic Bayesian Brain. In: Open Mind. Frankfurt am Main: MIND Group (2014)
37. Shannon, C.E.: Coding theorems for a discrete source with a fidelity criterion. IRE Nat. Conv. Rec **4**(142–163), 1 (1959)
38. Todorov, E.: General duality between optimal control and estimation. In: 47th IEEE Conference on Decision and Control, CDC 2008, pp. 4286–4292. IEEE (2008)
39. Wiener, N.: Cybernetics or Control and Communication in the Animal and the Machine, vol. 25. MIT press, Cambridge (1961)
40. Wolpert, D.M., Kawato, M.: Multiple paired forward and inverse models for motor control. Neural Netw. **11**(7–8), 1317–1329 (1998)
41. Yang, L., Iglesias, P.A.: Positive feedback may cause the biphasic response observed in the chemoattractant-induced response of Dictyostelium cells. Syst. Control Lett. **55**(4), 329–337 (2006)
42. Yi, T.-M., Huang, Y., Simon, M.I., Doyle, J.: Robust perfect adaptation in bacterial chemotaxis through integral feedback control. Proc. Natl. Acad. Sci. **97**(9), 4649–4653 (2000)

Isotopic Inheritance: A Topological Approach to Genotype Transfer

Olga Lukyanova$^{(\boxtimes)}$ and Oleg Nikitin

Computing Center, Far Eastern Branch of Russian Academy of Sciences,
Khabarovsk, Russia
ollukyan@gmail.com, olegioner@gmail.com

Abstract. Classic genetic algorithms (GA) incorporate a concept of variability in genotype to phenotype mapping only by the notion of generative encodings. In such cases, the phenotype depends not only on the pure values of genes but also on their interaction and influence of other factors like the environment. In case of generative encoding GA not exact phenotype of an individual is inherited but the isomorphic function to form the individual. The same information may lead to the formation of different phenotypes due to alterations in conditions or permutations in the gene sequence. We propose a mathematical definition of the abstract notion of genetic variability and a new approach to the problem of the genotype transition to a new generation called "isotopic inheritance". To this extent, we apply the notion of topological isotopy and use the branch of low-dimensional topology, known as braid theory to encode the values of a gene of arbitrary length into the genome and to propagate the isomorphic genotypes among the offspring. We illustrate the propositional variations of such encoding in different applications. Multiple knapsack problem was solved using the proposed approach.

Keywords: Genetic algorithm · Braid theory · Isotopy
Generative encoding · Isotopic genetic algorithms
Isotopic inheritance · Braid genes

1 Introduction

Same genetic code may lead to different phenotype outcomes during inheritance in biological systems. This is the basis of an important part of the genetic variability in natural populations with sexual and asexual reproduction [1].

In most genetic algorithms, genotype to phenotype mapping is described using direct encoding, which does not allow variability as a result of development. Therefore, the paper [2] justifies the need for developmental encoding. The use of unambiguous descriptions of phenotypes in genetic programs must be supplemented with algorithmic approaches that involve the variability of not only information in the genome, but also how the genome is decoded in the process of ontogeny.

© Springer Nature Switzerland AG 2018
P. Manoonpong et al. (Eds.): SAB 2018, LNAI 10994, pp. 27–38, 2018.
https://doi.org/10.1007/978-3-319-97628-0_3

Some genetic algorithms (GA) incorporate a concept of variability in geno-type to phenotype mapping by the notion of generative encodings. In such cases, the phenotype depends not only on the pure values of genes but also on their interaction and influence of other factors like the environment. In case of generative encoding GA not exact phenotype of an individual is inherited but the isomorphic function to form the individual. The same information may lead to the formation of different phenotypes due to alterations in conditions or permutations in the gene sequence.

Presented paper is aimed to investigate the possibility of a mathematical representation of individual variability in sexual and asexual reproduction, using the mathematical topology, in particular, the concept of braid isotopy. The goal of the work is also to find out how braid notations can be used to describe some of the practical problems of evolutionary optimization.

In the paper, a mathematical definition of the abstract notion of genetic variability is proposed. The notion of topological isotopy is applied and the branch of low-dimensional topology, known as braid theory, is used to encode the values of a gene of arbitrary length into the genome and to propagate the isomorphic genotypes among the offspring.

2   Phenotype Inheritance Strategies in Genetic Algorithms

2.1   Direct Genotype to Phenotype Mapping

In classical genetic algorithms, the process of changing the phenotype from generation to generation is determined by genetic operators: selection, recombination, mutation and inversion.

Selection is the implementation of the choice between candidates for breeding. Recombination is an operation in which two chromosomes exchange their parts. For example, 1100 & 1010 → 1110 & 1000. A mutation is a random change of one or more positions on the chromosome. For example, 1010011 → 1010001. Inversion is a change in the order of the bits in the chromosome or in its fragment. For example, 1100 → 0011.

In selection only those individuals can become parents, whose fitness value is not less than the threshold value. The most famous of the types of selection are tournament [3] and roulette [4] (proportional) selections. Thus, in the process of selection in classical genetic algorithms, the phenotype directly in one form or another passes from generation to generation in the form of mixed and distorted chromosomes of the most successful individuals. The operators of recombination, mutation and inversion are responsible for mixing and distortion.

In recombination the created offspring must inherit the gene information from both parents. There are discrete recombination and crossover. Discrete recombination corresponds to the exchange of genes between individuals.

Recombination of binary strings is commonly called a crossover. In single-point crossover, a point inside the chromosome (crossover point) is randomly determined, and in that point both chromosomes are divided into two parts and are exchanged. In a two-point crossover (and multi-point crossover in general), chromosomes are considered as cycles, which are formed by connecting the ends of a linear chromosome together. To replace a segment of one cycle segment of another cycle requires the selection of two crossover points.

In classical genetic algorithms, the phenotype is uniquely decoded from a mixed and distorted genotype of the previous generation, which is called direct encoding.

2.2   Approaches to Indirect Encoding

Indirect encoding uses a bio-inspired approach of phylogeny to ontogeny developmental mapping. The genotype encodes not the phenotype itself, but the rules for its construction. When decoding the genotype, these rules are applied in a certain sequence, as a result of which the phenotype is constructed.

As proposed by Hinton and Nowlan [5] learning and indirect mapping of genotype to phenotype may play important role in evolutionary optimization process. The information obtained during the interaction with the environment may affect agent's development and guide the evolution. Results of this seminal paper has affected the development of an area of evo-devo simulation. One of the ways to implement the indirect encoding as proposed by Hinton and Nowlan might be the alley variation and interaction. This interaction might be also affected the environmental factors. In fact, Hinton and Nowlan described the effects of gene expression during the development process.

The concept of genetic expression in genetic algorithms was developed in [6–9]. In particular, in [7], genetic expression depends on the indicators of the environment and regulates the development of the animat morphology. Ross in [10] proposed the concept of Lamarckian evolution to incorporate skills acquired during an agent's life into a genetic code. In the paper, this concept was distanced from biological basis, but the inclusion of the skills is possible through epigenetic processes. Algorithms based on epigenetic expression have shown their practical applicability in solving applied problems [8,9]. The study of developmental algorithms in morphogenesis is performed in [11], where the concept of cellular automata was applied for ontogeny under conditions of different environments. Also, this issue is considered in [12]. The works of [13,14] show that a combination of evolution and development leads to an improvement in the adaptability of agents.

Also, it is necessary to note the generative approach HyperNEAT, proposed in [15]. This approach involves the use of compositional pattern producing network (CPPN) to generate neural networks or agent morphologies.

Thus, in the field of genetic algorithms, there are a number of works that presuppose the notation of generative rules of ontogeny in the genome. All of them are more or less algorithmic and suggest an iterative implementation in the simulation process. The results of different developmental situations from the same genome can be perceived as an isotope set in relation to the original genome.

In Sect. 4 of this paper, we propose to apply the mathematical topology for modeling genetic isotopes, bypassing the process of probabilistic iterative development.

3   Biological Basis of Inheritance Variability

Biological evolution is not designed to increase the complexity of agents, but to allow maximally short transfer of flexible development rules for the adaptive behavior of agents in a complex environment [16]. Thus, inheritance of genes is a compression of information about the response of a phenotype to various environmental conditions. The process of inheritance of a phenotype from generation to generation is not a direct prescription, but the inheritance of similar variations of features. In this case, the same signs can be manifested through several generations, which can be considered as isotopes of the same phenotype. Phenotypes vary due to the gene modifications, environmental factors, allelic variation and interaction, or mitotic recombination [17–19].

Thus, the interaction of a gene and the environment is a process in which a phenotype is expressed on the basis of a certain genotype and the effect of a environmental factor [18]. Change in the phenotype can not be explained simply by adding the action of the genetic factor (variant of the gene) and the action of the environment.

The problem of the contribution of genotype-environmental interaction to variability has been discussed for a long time [20]. The two main sources of differences between animals in the population (genotype and environment) are closely related to one another and are in continuous interaction. Differences between carriers of different genotypes may not appear if there are no environmental factors that transform genotypic differences into phenotypic ones.

At the same time, allelic genes, or alleles, are different forms of the same gene. In a diploid cell there can only be two alleles of one gene, since there are only two homologous chromosomes. If in the diploid cell in homologous chromosomes are the same alleles, then in the cell there is one allele. In a haploid cell there is always only one allele. In total, the number of alleles of one gene in a population can be more than two. In such cases, they speak of multiple allelism. Being in one organism (cell), allelic genes interact with each other. The basic and simplest type of interaction is complete dominance, when one allele, manifesting itself, completely suppresses the manifestation of the other. Complete domination is observed when, for example, the color of pea seeds is manifested. This type of interaction of allelic genes as incomplete dominance is common. In this case, the dominant allele only partially suppresses the recessive allele. As a result, a certain intermediate value of the trait appears in the phenotype. Thus, the interaction between alleles can also determine the development of the phenotype from the genotype [17].

The variability of genes is also affected by the mitotic crossover [19]. It is a type of genetic recombination that can take place in somatic cells in mitotic divisions in both sexes and asexual organisms (for example, certain unicellular fungi in which the sexual process is not known). In the case of asexual organisms, mitotic recombination is an important source of evolutionary changes. In addition, mitotic recombination can lead to mosaic expression of recessive traits in a heterozygous individual.

Thus, in nature, the variability of a phenotype is determined by various factors of reproduction that come into play after the formation of DNA. In most genetic algorithms, phenomena of this kind are not reflected. A new mathematical approach to describing variability in the process of inheritance is proposed below.

4   Isotopic Inheritance Theory

To represent the phenomenon of variable phenotype transition from one generation to another, it is proposed to apply the mathematical notion of topological isotopy - the equivalence relation. Isotopy here is a metaphor for the process of transmission and formation of various phenotypes from the same genotype. To reflect the phenomenon of changing features it is convenient to use the low-dimensional topology and its part, such as the theory of braids.

4.1   Braid Isotopy

In [21] the definition of braid was given as a set of cross-linked strands. The most common way to represent a braid is to draw a diagram with crossed strands (Fig. 1). Braids resulting after the moves are equivalent (or isotope) to the original braid.

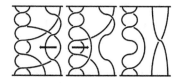

Fig. 1. Four stranded isotopic braids [22].

Thus, we call braids isotopic when we can continuously (without discontinuities and gluing) deform them in three-dimensional space into each other so that during the deformation all the strands of the braid go strictly downward. In particular, this means that the ends of the strands remain fixed, since they must move continuously on one side, and on the other hand they can occupy only a discrete set of positions. The braids have following rules: the strands are imagined to "flow" from the top to the bottom, the crossings happen at different vertical levels, and one strand is overlapped at each crossing (the lower strand passes behind the higher one in the real braid). For description of the braids, its diagrams are encoded into so-called braid words. It is a sequence of letters d_1^-,

d_1^+, d_2^-, d_2^+ etc. These letters specify braid diagrams (Fig. 2). We may encode which strand goes below, and which one passes over the other. Braid diagram with n strands has positions from 1 to n. If strand n is over crossed by strand $n+1$ it is coded by d_n^-. One calls d_1^- when the strand 2 crosses over the strand 1, d_2^- – when one strand (position 3) crosses over another (position 2) and so on. And vice versa: d_1^+ for strand 1 crosses over the strand 2 and d_2^+ for strand at position 2 crosses strand at 3.

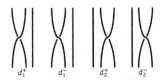

Fig. 2. Braid diagrams d_1^-, d_1^+, d_2^-, d_2^+ (adopted from [23]).

Braids can be used to encode any arbitrary alphabet and any text of arbitrary length. Below we propose how braids can be used to describe genes in genetic algorithms, and how their isotopes can model variable inheritance.

4.2   Braid Words for Genotype Encoding

In classical genetic algorithms, the problem can be written through location dependent encoding, where each parameter (gene) is written in its cell (location) (Fig. 3).

Gene 1	Gene 2	Gene 3
0 1 0	0 1 1	1 1 0

Fig. 3. Binary encoding for traditional genetic algorithms.

To encode individual genes, braid words can be used. This will allow us to compactly specify the parameters encoded by a separate gene. Such parameters can have any length. Also, the presence of several possible intersection alternatives (specified by the width of the braid n) allows to change the parameter on several levels. This can be used to describe gene expression, depending on several factors (Fig. 4). In addition, braids can be used to describe the implementation of the problem of any arbitrary complexity and length.

In Fig. 4, several braids describe several genes. An example of a genotype consisting of three genes (braids) is shown in the upper part of the figure. Each gene is noted by a sequence of intersections, each step in which can mean the interaction of a gene with an expression factor, describe the development of a phenotype or transmit parts of a feature description. Here is an example of the

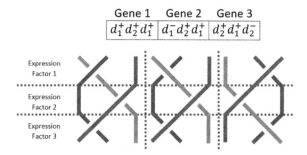

Fig. 4. Braid encoding for the genotype.

description of gene expression in several stages of ontogeny. Each level of the braid sets the realization of the phenotype at a certain stage. In this case, there are three stages of development of the phenotype in three separate features.

Figure 5 shows an example of specifying the structure of a deep neural network using braid genes. Two braids B1 – $d_1^+ d_2^+ d_1^+ d_3^- d_1^+ d_3^+$ and B2 – $d_2^+ d_1^+ d_2^+ d_1^+$ are presented. Here, using the same braid, two characteristics are described at once – the type of the layer of the neural network and the depth of the recurrent connection. The presence of an intersection on a vertical axis means that there is an appropriate variant of the trait. Three types of layers are defined: G – layer of gated recurrent units, C – convolutional layer, P – pooling layer. Three levels of the depth of the recurrent connection are also set: the link of the layer to itself is "0", the link of the layer to the previous layer is "−1", and the link of the layer through two layers back is "−2". They are shown in the diagrams of neural networks by red arrows. Each vertical level represents the implementation of a feature on a particular layer. Isotopic braid B2 has fewer intersections, so the neural network P2, which is phenotype mapping from B2, also has fewer layers. Since the braids B1 and B2 are isotopic, they can be represented as random different implementations of the same genotype. That is, B1 and B2 are a single object of heredity. Thus, the braid genes make it possible to describe the structure of deep neural networks with simple way of notation, and also indicate the correspondence between phenotypes using topological rules.

Another application of braid genes can be a description of the animat structure and the forms that takes its phenotype, depending on the development of the animat environment.

Figure 6 shows two genes, the limb number gene and the limb type gene. Two isotopes of each gene are also presented. For the number of limbs (Fig. 6A) we have $d_1^+ d_2^+ d_1^+ \rightarrow d_2^+ d_1^+ d_2^+$, for the type of limbs (Fig. 6B) – $d_1^- d_2^+ d_1^+ \rightarrow d_2^+ d_1^+ d_2^-$. The vertical intersections of each braid encode the number of limbs or the type of limbs, respectively, to the gene that the braid describes. The presence of negative and positive indices in braid words allows you to write $2n$ types of characteristics into the n-width braid. In Fig. 6A, the number of limbs are 1, 2, 4 and 6, respectively. In Fig. 6B shows different types of limbs: L - legs, F - fins, T - tentacles, W - weels. Negative indices define respectively one and six limbs, and legs and tentacles.

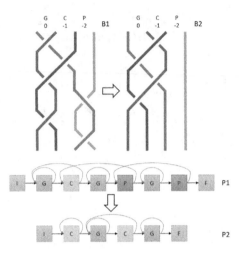

Fig. 5. Deep neural network described by the braid genotype.

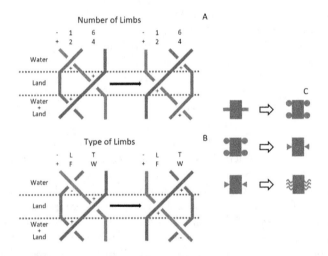

Fig. 6. An application of braid genes and their isotopes in the task of the animat evolution in various environments. A – two isotopes of genes responsible for the number of limbs, B – two isotopes of genes responsible for the type of limbs, C – the implementation of phenotypes in two isotopes and three types of environment.

Each vertical level of the braid indicates the type of environment in which the agent develops.

Thus, gene expression is determined, depending on the environment. Three types of environment are represented: water, land, and water mixed with land. The interaction of two genes and the environment generates one or another phenotype of the agent. Two isotopes of one genotype and their realization in three types of environment are described. In the cases presented in Fig. 6, four types of agents developed (Fig. 7). For example, in the case of a mixed environment (Water + Land), the first isotope took the form with two fins, and its isotope with six tentacles.

Thus, the description of the genotype with the help of braids allows describing both the procedure for constructing a complex phenotype (in the case of a deep neural network) and variants of phenotype development under different conditions (in the case of animats). The fact that braids allow the notation of any arbitrary length of text by any arbitrary alphabet makes it possible to describe more complex development situations, or morphologies in the problem posed.

4.3   Isotopic Inheritance

In the biological genome is not inherited a unique implementation of the phenotype, but a set of rules for the expression of the genotype in the phenotype, and we can talk about some randomness of the implementation of a phenotype, even in the case of asexual reproduction. A metaphor for such a random realization of a phenotype may be a topological isotopy. In this case, all isotopes are the results of the development of the same genotype. Thus, genotype is transmitted from generation to generation, but we can randomly choose any of its isotopes. This reflects phenomena such as the gene modifications, environmental factors, allelic variation and interaction, and mitotic recombination, which are accidentally manifested in the process of gene expression.

In turn, the mutation leads to a change in the initial notation of the genome. Thus, the mutated gene is not isotopic to the original. The notation of genes with the help of braids allows one to describe in a natural way the isotopic transition of genetic information from one generation to another. The same genome in different generations can lead to the emergence of various phenotypes, both as a consequence of environmental factors, and as a consequence of randomness in the realization of the phenotype.

In Fig. 6, you can see two isotopes of sets of genes that describe the phenotypes of the animat in different environments. In part C of the figure, shown changes in phenotypes due to the implementation of two variants of the genome isotopes. In the process of inheritance, such isotopes can be freely transformed into each other. If we accept that it is not the direct genome that inherits, but its isotope, then from generation to generation a transition from isotope 1 to isotope 2 and vice versa is possible, with the preservation of the hereditary connection. This will enrich the population with various gene implementations and expand the search space for optimal problem solving. This kind of indirect transmission of the embodiment of a phenotype from generation to generation can be called isotopic inheritance.

5   Practical Implementation

For testing the proposed approach against the real world tasks we applied it to the combinatorial optimization task of the packaging of a knapsack (Multiple Knapsack Problem) [24]. A formulation of the problem is stated below.

Fig. 7. Four types of animats described by braid genes.

Let's have n items and m backpacks ($n \leq m$). Each item has weight $p_j > 0$ and value $c_j > 0$ ($j = 1, 2, \cdots, n$), each backpack has its own capacity P_i at $i = 1, 2, \cdots, m$, $x_i \in 0, 1$. The task is to maximize $\sum_{i=1}^{m} \sum_{j=1}^{n} c_j x_{ij}$ so that the condition $\sum_{i=1}^{n} p_j x_{ij} \leq P_i$ is fulfilled for all $i = 1, 2, \cdots, n$, $\sum_{j=1}^{m} x_{ij} \leq 1$ for all $i = 1, 2, \cdots, n$.

We simulated the problem of a multiple knapsack. In this case, as a genotype there were a number of things, each of which has its own weight and utility. The task is to find a set so as to maximize utility, provided that the amount of backpacks is limited by weight. An example of a genotype and its isotopic representation is presented below: $[0, 0, 0, 3, 2, 1, 1, 0, 2, 0] \implies [3, 0, 0, 0, 2, 1, 0, 0, 1, 2]$, where numbers $1, 2, 3$ are numbers of the knapsacks.

Simulation experiments were conducted on the problem of the distribution of 5 things in 3 knapsacks. The size of the population was 20, the duration of the excitation was 100 iterations, inheritance was carried out by the roulette method, with a point crossover and a mutation probability of 0.1.

Comparison of the solution of the problem with and without isotopic inheritance has shown that the presence of an isomorphic mapping of a genotype into a phenotype leads to a more stable learning procedure (Fig. 8). Isotopic inheritance

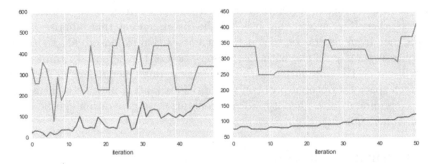

Fig. 8. Comparison of the results of a solution to the problem of a multiple knapsack without (left) and with (right) isotopic inheritance. The green line – the best result in the generation, the blue line – the average in the population. (Color figure online)

was introduced before the use of the mutation operator. In this case, the algorithm finds the optimal composition of the set of things more easily.

Thus, we can draw a conclusion about the practical applicability of the isotopic inheritance algorithm and also about the effectiveness of its work.

6   Conclusion

The article proposed a new approach to genetic coding, using braid theory, which reflects the biological aspects of genetic variability not related to the crossover and mutation, and should enrich the search space of genetic algorithms. Several examples of the application of this approach to real problems of evolutionary optimization are described. It was shown that isomorphic inheritance improves the performance of GA on a simple task of the knapsack problem. The inheritance approach stabilized learning and enriched the search space.

Reducing the direct connection between the genotype and the phenotype, in consequence of the phenomenon of isotopic inheritance described in Sects. 4.1 and 4.2 of the paper, may lead to a loss of direction in the search for a genetic algorithm. This question requires a separate study.

Nevertheless, the encoding of the genome, through the braid genes, makes it possible to describe a wide range of problems compactly and conveniently, and the natural properties of topological braids such as isotopy create possibilities for interesting experiments in the field of inheritance variability.

Acknowledgments. The research was funded by RFBR project No. 18-31-00188. The calculations were carried out using methods and technologies which development was funded by the grant of the Program of Basic Research FEB RAS "Far East" to the research projects No. 18-5-100. The computing resources of the Shared Facility Center "Data Center of FEB RAS" (Khabarovsk) were used to carry out calculations. We thank Alexey Kunin for technical assistance in the preparation of the manuscript.

References

1. Bengtsson, B.O.: Genetic variation in organisms with sexual and asexual reproduction. J. Evol. Biol. **16**, 189–199 (2003)
2. Pigliucci, M.: Genotype-phenotype mapping and the end of the 'genes as blueprint' metaphor. Phil. Trans. R. Soc. B. **365**, 557–566 (2010)
3. Brindle, A.: Genetic algorithms for function optimization (Doctoral dissertation and Technical report TR81-2). University of Alberta, Edmonton (1981)
4. Holland, J.H.: Adaptation in Natural and Artificial Systems. University of Michigan Press, Ann Arbor (1975)
5. Hinton, G.E., Nowlan, S.J.: How learning can guide evolution. Complex Syst. **1**(3), 495–502 (1987)
6. Kargupta, H.: The gene expression messy genetic algorithm. In: Proceedings of IEEE International Conference on Evolutionary Computation, pp. 814–819, Nagoya (1996)

7. Eggenberger, P.: Evolving morphologies of simulated 3D organisms based on differential gene expression. In: Husbands, P., Harvey, I. (eds.) Proceedings of the Fourth European Conference on Artificial Life, pp. 205–213. MIT Press, Cambridge (1997)

8. Ferreira, C.: Gene expression programming: a new adaptive algorithm for solving problems. Complex Syst. **13**(2), 87–129 (2001)

9. Wu, A.S., Garibay, I.: The proportional genetic algorithm: gene expression in a genetic algorithm. Genetic Program. Evolvable Mach. **3**(2), 157–192 (2002)

10. Ross, B.: A Lamarckian evolution strategy for genetic algorithms. In: Chambers, L.D. (ed.) Practical Handbook of Genetic Algorithms: Complex Coding Systems, vol. 3, pp. 1–16. CRC Press, Boca Raton (1999)

11. Kowaliw, T., Grogono, P., Kharma, N.: Bluenome: a novel developmental model of artificial morphogenesis. In: Deb, K. (ed.) GECCO 2004. LNCS, vol. 3102, pp. 93–104. Springer, Heidelberg (2004). https://doi.org/10.1007/978-3-540-24854-5_9

12. Auerbach, J.E., Bongard, J.C.: Environmental influence on the evolution of morphological complexity in machines. PLOS Comput. Biol. **10**(1), e1003399 (2014)

13. Kriegman, S., Cheney, N., Corucci, F., Bongard, J.C.: A minimal developmental model can increase evolvability in soft robots. In: Proceedings of GECCO 2017, pp. 131–138. ACM, New York (2017)

14. Veenstra, F., Faina, A., Risi, S., Stoy, K.: Evolution and morphogenesis of simulated modular robots: a comparison between a direct and generative encoding. In: Squillero, G., Sim, K. (eds.) EvoApplications 2017. LNCS, vol. 10199, pp. 870–885. Springer, Cham (2017). https://doi.org/10.1007/978-3-319-55849-3_56

15. Stanley, K.O., DAmbrosio, D.B., Gauci, J.: A hypercube-based indirect encoding for evolving largescale neural networks. Artif. Life **15**(2), 185–212 (2009)

16. Wolfram, S.: A New Kind of Science. Wolfram Media, Champaign (2002)

17. Lo, H.S., et al.: Allelic variation in gene expression is common in the human genome. Genome Res. **13**(8), 1855–1862 (2003)

18. Lobo, I.: Same genetic mutation, different genetic disease phenotype. Nature Educ. **1**(1), 64 (2008)

19. LaFave, M.C., Sekelsky, J.: Mitotic recombination: why? when? how? where? PLoS Genet. **5**(3), e1000411 (2009)

20. Haldane, J.B.S.: The interaction of nature and nurture. Ann. Eugen. **13**, 197–202 (1946)

21. Lukyanova, O., Nikitin, O.: Neuronal topology as set of braids: information processing, transformation and dynamics. Opt. Memory Neural Netw. (Inf. Optics) **26**(3), 172–181 (2017)

22. Sossinsky, A.: Knots, Mathematics with a Twist. Harvard University Press, Cambridge (2002)

23. Kassel, C., Turaev, V.: Braid Groups. Graduate Texts in Mathematics. Springer, New York (2008). https://doi.org/10.1007/978-0-387-68548-9

24. Martello, S., Toth, P.: Knapsack Problems: Algorithms and Computer Implementations. Wiley, New York (1990)

Perception and Motor Control

Towards Rich Motion Skills
with the Lightweight Quadruped Robot
Serval - A Design, Control
and Experimental Study

Peter Eckert[1]([✉]), Anja E. M. Schmerbauch[1,2], Tomislav Horvat[1],
Katja Söhnel[3], Martin S. Fischer[3], Hartmut Witte[2], and Auke J. Ijspeert[1]

[1] Biorobotics Laboratory, École Polytechnique Fédérale de Lausanne,
Lausanne, Switzerland
peter.eckert@epfl.ch
[2] Fachgebiet Biomechatronik, Technische Universität Ilmenau, Ilmenau, Germany
[3] Institut für Zoologie und Evolutionsforschung, Friedrich-Schiller-Universität Jena,
Jena, Germany

Abstract. Bio-inspired robotic designs introducing and benefiting from morphological aspects present in animals allowed the generation of fast, robust and energy efficient locomotion. We used engineering tools and interdisciplinary knowledge transferred from biology to build low-cost robots able to achieve a certain level of versatility. Serval, a compliant quadruped robot with actuated spine and high range of motion in all joints was developed to address the question of what mechatronic complexity is needed to achieve rich motion skills. In our experiments, the robot presented a high level of versatility (number of skills) at medium speed, with a minimal control effort and, in this article, no usage of its spine. Implementing a basic kinematics-duplication from dogs, we found strengths to emphasize, weaknesses to correct and made Serval ready for future attempts to achieve more agile locomotion. In particular, we investigated the following skills: trot, bound (crouched), sidestep, turn with a radius, ascend slopes including flat ground transition, perform single and double step-downs, fall, trot over bumpy terrain, lie/sit down, and stand up.

1 Introduction

When we think about animals, we see many species moving dynamically in their natural habitats. While one also observes static behaviors, being in motion is characteristic for most of animal life. The variety of motion skills present in a single species is vast and more refined than what robotics researchers were able to achieve so far with technology. Two aspects that are especially striking in this context are versatility and agility. Presented originally in [1], agility is defined as follows: *"Agility is representing a previously acquired and size dependent set of locomotion skills, executed in a precise, fast and ideally reflexive manner to an*

© Springer Nature Switzerland AG 2018
P. Manoonpong et al. (Eds.): SAB 2018, LNAI 10994, pp. 41–55, 2018.
https://doi.org/10.1007/978-3-319-97628-0_4

outside stimulus.". Comparing this definition to the one of versatility from the Oxford Dictionary (en.oxforddictionaries.com): *"Ability to adapt or be adapted to many functions or activities"* a new definition of agility as **fast versatility** or **fast execution of rich motion skills** may be called for. In consequence, one way to reach agility, could be to first achieve versatile behavior followed by an increase in its execution speed.

Machines with relatively simple underlying principles (e.g., cars or bikes) can move very well in our environment and navigate even through difficult terrains. In legged robotics on the other hand, whose motivation is high adaptability to uneven or difficult terrains [2], such fast and reliable locomotion is yet to be achieved. It is still unclear, although explored in many laboratories all over the world, what level of mechatronic complexity is minimally needed and sufficient to realize agile motion.

Two approaches, often described and used in control, are templates and anchors [3], both incorporating information gained from observation and analysis of animals. Hereby, a template, following the strictly bio-inspired direction, is simplifying the animal and its motion to the highest degree, enabling comparison on the principle-level between species. Template models can be tested against empirical data (for example the spring-loaded-inverted-pendulum-model). Anchors build upon templates and embed them in a more complex and realistic morphological and physiological model (towards bio-mimicry). Here details ranging from muscle-placement, specific joint torques up to the underlying neural control networks can and should be integrated. Usage of both, templates and anchors, in combination with detailed mechanical models and real robot hardware as 'physical simulator' to explore specific (neuro-)mechanical questions is possible. Our work uses such approaches, but in a middle way, finding an acceptable level between biological detail and complexity of implementation. This includes simple, template-like control methods with slight bio-mimicry influences, e.g., in geometry and kinematics of the robot.

This article, based on a review of the versatile behavior found in legged locomotion, will highlight our efforts in creating a small and safe experimental platform achieving the first step towards agility and rich motion skills (only a subset of implemented motions is described here, due to the scope of the article). The purpose of the robot is its usage as a system for active research, as well as an educational tool.

2   State-of-the-Art: Versatile Legged, Terrestrial Robots

Terrestrial legged robots range from insect scale up to large systems the size of a small horse [4]. In this short review, our focus on versatility shown by different quadrupedal robots is highlighted in the following paragraphs. Classifying existing robots by their versatility or agility in a benchmark like [1] is difficult, as only information from previous publications is available. Often included are however the absolute or relative speed, geometric measures and sometimes slope-inclinations, jump-heights or the robot's capabilities for turning.

We group robots according to their mass (Group 1: $m > 10\,kg$, Group 2: $10\,kg > m$), presenting them on a qualitative level for their versatility and main characteristics.

In the **first group**, consisting of MIT-Cheetah I+II [5,6], ANYmal [7], StarlETH [8], Scout II [9] and Canid [10], the use of electrical actuation is prominent. The robots' size and weight are such that one can employ customized electric actuation with or without passive series elastic elements, e.g., high-power-density in MIT-Cheetah and highly integrated series elastic actuators in ANYmal. This group of robots is often adapted to different purposes reaching from navigation and spatial mapping in cluttered terrains, mobile manipulation to very high dynamic locomotion. The available payload allows for a high variety of sensors to be equipped and used in model-based, closed loop control schemes. Another important aspect is the possibility to handle the robot with less than two handlers, making them very well suited as sturdy experimental platforms, also for questions other than locomotion. Restrictions for morphological research is present due to the weight. Filigrane passive compliant structures, like toes, or a partially passive compliant spine are complicated to implement with current technical means, as the employable compliance can often not support the robot's mass. This scaling related effect, is most visible in MIT-Cheetah, switching from an actively bendable spine in Version I to a completely rigid trunk in Version 2. An exception, due to its weight of roughly $11\,kg$, is presented with Canid. Its flexible trunk and wheg-like legs enable the robot to jump, indicating the use of high power density, but commercially standard electric actuation.

Small robots, here even under $5\,kg$, represent the **second** and most influential group for the work presented in this paper. We selected Tekken 1 and 2 [11,12], Puppy I and II [13,14] as well as Cheetah-Cub [15], Bobcat [16], Oncilla [17], Lynx [18] and Cheetah-Cub-S [19]. For this class of robots a different development scheme can be employed. Additional to a general light-weight approach, very high cost-reduction becomes feasible. As the robots weight decreases, so does the necessary torque to induce movement, enabling the use of purely commercial actuation technologies down to high grade RC-servo-motors. Most robots presented, use passive elastic elements in legs or trunk to minimize active actuation and research more morphology-related aspects of locomotion. This includes the impact of the spring loaded pantographic leg-structures inspired by [20] and control methods (like CPGs) not relying on precise sensor feedback or torque-control. Nevertheless, as shown already in Tekken 1 and 2, stable locomotion, even on relatively rough terrain remains possible. This ability is refined by adding feedback in form of different reflexes to adapt and react to the environment, as in the Tekken robots and Oncilla [2]. Losing the capability to carry heavy sensory equipment, e.g. LIDAR (needed for quick and precise spatial mapping) is compensated with an ease of use as experimental platforms for template research, e.g. the application of flexible spines for steering and improved locomotion (Cheetah-Cub-S, Lynx). Small robots can be handled safely by a single operator, even after very basic training. Cost-efficient production allows for groups without high budget to copy, maintain and use small robots as physical simulators and thus increases options to verify theoretical or simulation work.

The presented state-of-the-art clarified for us two main aspects: (a) Visible from our qualitative analysis we can state that researching versatility and often agility is possible with any size of quadrupedal robot. The distinction has to be made if one desires an all-round robot (group 1) or a more specialized system (group 2, besides Oncilla). (b) Our intent to use robots as educational and research platforms, implies safety and ease of use to be of the essence. This is generally prominent group 2. For Serval, we aim for a middle way in between both groups.

3   Robot Development

Animals, despite their enormous skill to balance, often trip, fall or run into their surroundings. To compensate for the lack of environmental sensing and reflexive mechanisms, our robot needed at least a sturdiness somewhat close to the animal one. Compliant elements in key positions in combination with a sturdy, but very lightweight skeleton should thus build Serval's mechanical core, also resulting in a low inertia system. Relatively high power-density actuation and processing power enable fast execution of motion commands. If these measures fail to enhance the movement and protect our robot, the modular design approach enables easy and quick repairs. Following our goal of reaching versatile movement, our robot control needed to consist of a flexible and modular approach. Formed around an open-loop CPG-controller for basic movement generation, we implemented different behaviors, which cohere with our vision of versatility and can potentially be executed quickly to achieve agile movement. If these aspects work together symbiotically within a compliant and relatively powerful mechatronic design, we are confident to be capable of a good grade of versatility or even agility.

3.1   Mechanical Development

Mechanical development in Serval presents a combination of tested mechanisms from previous robots with the goal of enhancing their advantages while canceling out as many disadvantages as possible. The resulting robot consisted of a modular design built around Dynamixel MX64R/MX28R servo motors and an Odroid XU4 SBC. One can distinguish three reusable main units: (1) trunk, (2) leg and (3) spine unit, illustrated in Fig. 1. The trunk-unit acts mainly as a connection hub, and housing for the legs, spine as well as electronics. Due to the scope of the paper, we will refrain from describing it in detail. The units are integrated and extended with a new foot design as well as in-series elastics for motor and mechanics protection from impacts. To reach a lightweight as possible design and with ease of implementation in mind, we built the robot's skeleton from lightweight aluminum (Al), steel (only for axes) and POM (engineering plastic), machined with classical CNC-milling, CNC-Laser, and bending techniques. The leg-unit incorporated at its base an ASLP-leg (advanced spring-loaded pantograph) in the design of the previously developed Cheetah-Cub-AL (go.epfl.ch/CheetahCubAL), see Fig. 2. Due to the size of the robot and thus the legs' dimensions and resulting lever arms, relatively stiff springs had to be integrated. Please find a 3D-PDF of Serval here: go.epfl.ch/3DPDFServal.

Leg-Unit: ASLP segmentation (fore and hind differ) was kept as a scaled version of Cheetah-Cub-AL to re-use as much of the previous design as possible. Additionally, a passive-compliant carpal-joint (wrist joint) was added to the fore legs, to test the possibility of small-step-ups without sensory feedback (future work). The leg unit was iterated once, as the diagonal spring-mechanism caused failure of blocking the leg due to quick wear and tear of its rectangular guidance (due to manufacturing imprecision and choice of material). After exchanging it with precisely turned AL-inner- and POM-outer-guide, this issue was solved. The legs' springs work as an antagonist to their respective motor, thus enabling relatively fast extension with a force that can be deduced from the springs' prior deflection. Although this fast extension might lead towards explosive movements in the future, a trade-off had to be chosen in the current design. For motions like jumping, high forces are needed, thus depicting the need for very high spring-stiffness. This would exceeding the resulting compression capabilities of the servo-motors employed in our work and limit the adaptability of the robot to uneven terrains of falls due to a lack of energy absorption/dissipation in the legs. Jumping would thus only be possible with a different actuator choice or a supplementary leg extension mechanism. Between robot adduction/abduction (AA), with

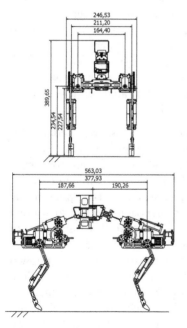

Fig. 1. Front and side view of Serval with characteristic measurements, abstracted CAD.

Table 1. Characteristic values of Serval; Geometric measures extracted from CAD.

	Unit	Serval
Height: Ground-Hip	[mm]	228
Width: Leg-leg	[mm]	211
Length: Hip-Hip	[mm]	378
Mass: Total	[g]	3560
Mass: Electronics	[g]	2167
Mass: Mechanics	[g]	1393
Stiffness: Diagonal Spring	[N/mm]	7.76
Stiffness: Parallel Spring	[N/mm]	9.06
Stiffness: Foot Spring	[N/mm]	1.98 (x2)
Stiffness: Adduction/ Abduction	[Nm/rad]	253.2
Stiffness: Spine	[N/mm]	8.4/ 52
DOF: Actuated		15
ROM: Fore Hip	[°]	+76/-50
ROM: Hind Hip	[°]	+84/-64
ROM: Knee	[mm]	93
ROM: Adduction/ Abduction	[°]	+90/-70
ROM: Spine	[°]	±90/±30
Motor: Servo		Dx MX28R/64R
Voltage: Servo	[V]	10-14.8
Stall torgue: Servo	[Nm]	2.5/ 6 (12V)
No load speed: Servo	[°/s]	330/ 378 (12V)
Gear ratio: Servo		193:1/ 200:1
Single Board Computer		Odroid XU4
LiPo-Battery		3S-3.3Ah-25C
Iterations		1.5

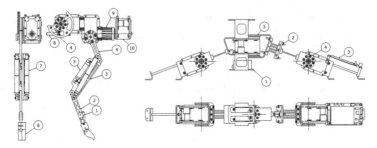

Fig. 2. (left) Serval's leg unit from a side and front view: (1) carpal joint, (2) l3-segment (parallel to l1), (3) l2-segment, (4) l1-segment (parallel to l3), (5) parallel spring (uncompressed), (6) compliant foot (2 toes), (7) diagonal springs (symmetric to saggital plane of the leg), (8) rotary fixation for the leg-unit, (9) AA-in-series-elastic, (10) AA-motor with axis slightly displace from hip axis, higher ground-clearance when moving to the outside, smaller to the inside; (right) Serval's spine unit from a side and top view: (1) IMU fixation, (2) Cross-joint with Nitinol-leafsprings, (3) screen for basic HMI, (4) up-down DOF, (5) steering DOF.

direct actuation located roughly on the hip axis height, an in-series elastic was mounted. The very stiff mechanism, still used in the experiments presented in this paper, was later-on exchanged for a torsion tube-like structure, made from NiTinol-wires in its super-elastic state. Flexion of the leg due to external forces resulted in a torsional displacement, reducing direct impact propagation to the AA-actuation. This mechanism could be combined with a rotational damper to dissipate impact energy instead of just smoothing the peak forces. After testing different designs, we decided to include a segmented, spring-loaded foot with two rounded, claw-shaped toes. We hypothesized the need for ground adaptation due to the large AA-capability of the robot (changing lateral angle to the ground) and expected better grip on rough or granular terrains.

Spine-Unit: Three active DOF, one for rotation in the transversal and two for rotation in the saggital plane, were forming the core of the spine including a small handle and IMU-connector on the middle motor. The elements were connected through bent AL-pieces and leaf-springs made of four NiTinol wires in parallel. The hind elastics were connected in a cross shape, stiffening the spine in one direction and enabling compliant behavior in the orthogonal other to comply with the in-series load for their respective motor. The overall arrangement of the springs is approximating a classical rectangular leaf spring. This results in different overall behavior of the springs depending on the direction of the applied forces, inducing small displacement in one and larger in the orthogonal direction. For adjustment of the allowable deflection and stiffness of the springs, clamps can be added to the spring-fixations, shortening the free length of the mechanism. Stiffness is also reducable by removing wires from the set. For our spine, we desire relatively stiff connections enabling the direct transmission of forces in the steering direction and softer springs for saggital movement (Table 1).

We believe that a certain amount of compliance in the spine is necessary when impacts are too large to be absorbed by the legs alone and is yet to be tested. Usage of NiTinol as shape memory allow with resulting stiffness changes was omitted due slow reaction time (heating/cooling) and high unpredictability in its transition states.

3.2   Control Development

The first implementation towards versatile movement with Serval consisted of replaying modified kinematic data from dogs [21]. This approach, due to the readiness-state of the robot hardware and control (no sensors integrated at the time, no reflexes), was performed in open-loop and is thus a basic control to be extended in current and future work.

Analysis and mathematical representation of motion capture (MOCAP) data from trained Border Collies enabled the foot trajectory generation for Serval's inverse kinematics control. Four dogs' data was available to be processed to obtain kinematic data of different gaits. Figure 3 shows an example for fore- and hind-foot trajectory, illustrating differences in, e.g., vertical displacement as well as distances to hip joint axes. All Border Collies were taller than 45 cm at withers resulting in the need for scaling of established trajectories to the robot's size. We fitted four cubic Bézier curves to the data, to mathematically recreate the complex shape of a real animal foot-locus. Junction Points were positioned vertically to the hip axis and on the transition from stance to swing and swing to stance phase. Consequently, the calculation of the inner Bézier points completed the definition the dog foot-loci with cubic Bézier curves. We proposed a parametrization approach using hip height (H) as origin (x_0, z_0), step height (h), compression factor (c) included in $(h' = (1 - c)h)$, step length (SL) and length proportion per direction (LR and LL) for trajectory modification in experiments, see Fig. 3. The use of take-off and touchdown angles, like in Cheetah-Cub-AL's foot-locus parametrization [22], was deliberately omitted to keep the ratios and proportions of foot trajectories imported from animal data intact but retain the

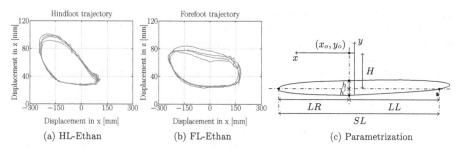

(a) HL-Ethan   (b) FL-Ethan   (c) Parametrization

Fig. 3. (a) and (b) Forefoot and hindfoot trajectory for trot of a Border Collie (Ethan); head to the right; rather flat elliptic shape in the fore and angled elliptic shape in the hind (more ground clearance); (c) Parametrization method for experimental implementation.

possibility for adjustment of the general trajectory size. Having extracted the foot-loci from cubic Bézier curve interpolation (and depending on the leg timing), the data was ready to be "replayed" by Serval. Underlying control was using an adapted framework from Pleurobot [23]. On a high-level, we used the controller's state machine along with a CPG-network to update the foot-position constantly, generating our different motions. Duty factor was not commanded explicitly but depended on the implemented trajectory. The combination of the trajectory with correct timing of the inter-limb coordination let to a specific gait.

4 Experimental Validation

Most of the here presented experiments do not yet employ active trunk movement. For debugging purposes, we decided to block the spine with two POM plates and free it only for use in turning maneuvers. All tests presented here were done tethered for off-board power supply.

After preliminary adaptation of the scaling to match our dog-data to the robot's geometry, we focused on different skills to test robot mechanics and its suitability for versatile and agile locomotion. A subset is presented in this article: trot (with and without AA), bound (crouched), sidestepping, turning with

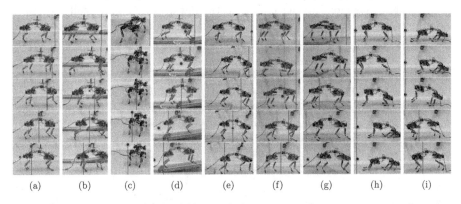

(a)   (b)   (c)   (d)   (e)   (f)   (g)   (h)   (i)

Fig. 4. Snapshots of Serval (one skill per column, start on the top, transparent bar gives a fixed reference between images: (a) trot (dynamic movement, sliding at touch-down and toe-off, decreasing efficiency, speed at 0.83 m/s), (b) crouched bound (dynamic movement, left-right-symmetry, sliding prominent at toe-off, small turn to the left), (c) side-step (lateral combined with backwards motion, stick-slip and stuttering due to foot-geometry and small ground clearance during swing), (d) bounding upslope (16° inclination (≈28.7%), transition from flat to inclined terrain, heading relatively straight), (e) trotting down a single step (63 mm (≈30% of leg length), success-rate of 70%, shock absorption through the parallel spring), (f) falling (height ≈70% of leg length while trotting, success rate >90%, complete flexion of the parallel spring and touch down of the knee, passive flexion of the diagonal springs), (g) trotting on a smoothed bumpy terrain (after initial step-down, deviations from straight path and uncontrolled movements are present), (h) and (i) lying down and standing up (fully hard-coded motion-sequence inspired by dogs).

a radius, slope-up with a flat ground transition, single and double step-down, fall absorption, rough terrain, lying/sitting down, and standing up (Videos: go.epfl.ch/ExperimentsServal[1]).

4.1   Flat Terrain Locomotion

Trot: Following the gaits often observed in animals, the running trot was tested ($h = 0.03$ m, $SL = 0.15$ m and $F = 1.5$ Hz), see Fig. 4 for snapshots. We used two different settings for the AA. When moving on flat ground, the hind legs were flexed towards the sagittal plane and forelegs extended in the opposite direction. This posture is observed in dogs when moving in medium to high speeds to possibly enable overlapping of their feet during motion. This way, the hind legs can provide most of the propulsion whereas fore legs stabilize the robot. For other tasks, like step-downs or backward trot, we set the AA straight. This decreased variation in the robot's roll-angle when perturbed and was thus useful in cases where self-stabilization was the top priority. In Fig. 5 we present an example of GRFs (Kistler, type 9260AA3, mounted side-by-side (left-right) and covered with non-reflective tape within a catwalk) that mainly confirms the visual impression of characteristic trotting and bounding from Fig. 4. The main challenge in obtaining reliable GRF-data was the feed-forward generation of the robot gait. The robot had to run straight over the two force-plates, allowing the distinction between left and right feet and thus the generation of valid GRF-gait-data. As the robot was often reacting to changing surface conditions by changes in direction, separated touchdowns were not impossible, but difficult to achieve. This increased the number of experimental runs that had to be performed to acquire clean GRF-data. The robot showed the main characteristics of a trot-pattern repetitively, see Fig. 4a.

The reduced controllability of the ASLP leg due to under-actuation can explain the phenomenon of occurring slippage to compensate early touch down of the feet, already observed in Oncilla and Cheetah-Cub [2,15,17]. Further, as we were using a gait not tailored specifically to the robot but stemming from kinematic recordings, a mismatch is possible. From a mechanics point of view, we see future need for new materials with anisotropic friction to enhance propulsion in one and allow for slippage in the other direction. Nevertheless, the trot gait was very stable, out-of-the-box and enabled the robot to locomote at a speed of 0.83 m/s (FR = 0.32) with a visible trot footfall pattern, see Fig. 5a.

Bound (Crouched): The crouched bound (Fig. 4b), is used by cats to climb very steep slopes over 50% inclination [24], but is also a useful gait when testing active spine movement [16,18]. For us, as we did not yet free the spine, the crouched bound was used mainly for slopes, as described later. Nevertheless, a feasible gait was also obtained when running on flat terrain. Walking foot-loci with changed inter-limb timing were used to achieve the motion. Bounding showed good directional stability, visible from Fig. 4 and an almost perfect

---

[1] No sound available due to recording without microphone.

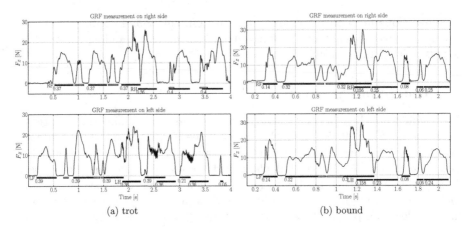

Fig. 5. (a) GRF measurements for a trot gait, footfall patterns indicated in black with expected characteristic appearance, foot sliding is represented by white boxes, and individual duty factors are marked, mean duty factors: $DF_{LF} = 0.36$, $DF_{LH} = 0.37$ $DF_{RF} = 0.37$, $DF_{RH} = 0.39$ and $DF_{av} = 0.37$; GRF patterns are similar to Cheetah-Cub, with never the full robot weight (33 N, without battery) on one single foot; Peaks are visible, e.g., at 1 s on the left fore foot that was caused by the hind foot obtaining high traction leading to higher compensation forces in the diagonally opposite foot; (b) GRF measurements for a bounding gait using a crouched posture; mean duty factors: $DF_{LF} = 0.38$, $DF_{LH} = 0.34$ $DF_{RF} = 0.39$, $DF_{RH} = 0.36$ and $DF_{av} = 0.37$; robot weight is evenly distributed on two sagitally opposite feet (left-right-symmetry); the peak at ≈ 1.3 s is an example of all feet touching the ground at the same time, resulting in application of the robot's full weight as vertical force.

representation of the desired footfall pattern, illustrated in Fig. 5b. At seldom occasions stick-slip is visible. We hypothesize that improvement towards a non-slip gait is possible when freeing the spine and improving ground contact with anisotropic friction material on the feet.

Side-Stepping: We included lateral side-stepping in our initial experiments as preparation for later execution of a lateral stepping reflex [2]. A spatial 8-figures was commanded to the robot's feet, hence using the AA to push the robot to one side and shifting body weight away from the side whose feet should be in swing-phase. Without touching the ground, Serval was able to perform the task, but as soon as in contact, stick-slip with the feet's hard edges due to little ground clearance made a movement impossible. The snapshots in Fig. 4 demonstrate an alternative, artificial gait using AA, allowing for lateral-aft motion. A pure lateral movement might be difficult to achieve, if posture is not kept balanced through a posture-adaptation-reflex, allowing the swing legs to execute their movement without touching the ground.

Turning with a Radius: The last movement on flat ground essential for a versatile system is the ability to turn. Here we used a combination of asymmetric stride length [17] and lateral spine-deflection. The resulting turn was again rather perturbed by slippage but reaching a small turning-radius of 0.58 m. Adding the AA-movement to the two previously mentioned strategies should reduce slippage and make turning more repeatable.

4.2 Inclined Surfaces, Perturbations, Stability, and Rest-Position

Slope-Up: Experiments have been performed to identify the maximum slope Serval can climb up, using a bound and crouched posture adaptation, see Fig. 4d. The maximum inclination feasible with an open-loop gait amounted to 20° (≈36.4%) with a transition from flat ground to the slope. Without heading-correction, substantial slippage in the propulsion phase of the toe-off and drift to the side could be observed. Serval could repeatedly move on a 16° (≈28.7%) slope with little drift. Smaller inclinations could also be achieved with a trot-gait. Locomotion down-slope was also possible. We did no test lateral inclinations, as an appropriate test setup was not available at the time.

Single and Double Step-Down: We included step-down experiments to demonstrate the self-stabilizing behavior of the robot and the gait robustness, see Fig. 4e. The goal was the determination of the maximum step height which the robot can go down in open-loop while reliably using its legs' compliance. The applied gait was an unmodified trot. The requirements for a successful try was the continuation of a stable gait for at least 2 m after step-down. We performed at least ten runs per step-height. With reliability of 100% Serval adapted to step-downs of 53 mm that amounts to ≈25.2% of its leg length. The largest step of 63 mm (≈30% of leg length) resulted in a success-rate of 70%. Step downs over two consecutive steps were successfully performed in 90% of the cases, at the height of 26 mm (≈12.5% of leg length).

Fall Absorption: Robustness is key, as versatile and agile motion can lead to falls and failures rather quickly. Dropping the robot from a maximum height of 70% leg length while running with a trot showed that it was possible to overcome impacts and continue locomotion in this idealized scenario, see Fig. 4f. In the first images after touch down one notices the strong deflection of all leg-springs and the resulting push-off. A critical point is presented by the touch-down of the leg's middle segment and full compression of the parallel spring already for this drop height. However, the robot was always able to regain a stable trot after some stepping cycles. This result is encouraging, as it shows robustness as long as the force is transmitted in a way that the leg-compliance can disperse the impact. Additional tests from other angles and heights should be performed to characterize the robot further and test the new in-series implementation in AA and spine joints.

Rough Terrain: Moving over sharp vertical obstacles, using only open-loop, was already found to be widely impossible in our experiments with Oncilla [2]. As an additional test of Serval's stability, we decided to let it run on a smooth, but bumpy terrain after a small step-down, see Fig. 4g. Without controlling the heading, the robot was sliding to different sides, moving backward, but in the end, finishing the distance over the plate. This highly irregular behavior is of limited use in real scenarios, but again, underlines the robot's stability due to its compliance. Heading and posture control may build on this stability to enable new application environments and increase the robot's real-world capacity and adaptation.

Lying/Sitting Down and Standing-Up: We defined the transition from sitting/lying to a normal standing posture as a part of versatility. Consequently, these behaviors were implemented and tested. Kinematic data for the robot joints was extracted from MOCAP of Border Collies, and the movements were implemented with success, as hard-coded motion-sequences [25], see Fig. 4h, i.

5   Conclusion and Discussion

When comparing to our reviewed state-of-the-art, Serval is positioning itself as a versatile robot with a high level of mobility at medium speeds. Robots showing comparable skills are Oncilla, Tekken 2 and Scout II. ANYmal, StarlETH, Canid, as well as the MIT Cheetahs are still ahead regarding performance (through higher grade actuation and closed-loop control) or simply present a different set of skills (such as walking stairs or jumping). With the number of successfully implemented skills, using a basic kinematics-duplication, we debugged the robot hardware, found out strengths to emphasize (compliance and adaptable feet), and weaknesses to correct (friction of ground contact and low stiffness of spine/AA). All in all the initial tests shown here were a great success generating valuable insights towards hard- and software development.

Our study of **flat terrain locomotion** demonstrated Serval's ability to achieve many motion patterns, simply by replaying parameterized dog-foot-loci. Further investigation of optimal gaits tailored to the robot should improve the existing patterns. Further work should investigate if bio-inspired, but artificially generated trajectories, like in the Cheetah-Cub-Family and Oncilla, have an advantage over replaying kinematic data from example animals. The robot's capability of **ascending slopes** in open-loop is very promising. It already improved markedly from max 15° in Oncilla (closed-loop) to max 20° in Serval. In further work and with PAD as well as better surface friction included, even steeper slopes will be feasible. Regarding **stepping down**, Serval showed remarkable results and thus followed up on the success in Cheetah-Cub, even increasing the percentile maximal step height by 10% and success-ratio by 50% in this direct comparison [15]. The next logical step is to increase maneuverability by implementing step-ups through reflexes. Both **rough terrain** locomotion as well as **fall absorption** were handled repeatedly well. Aforementioned skill is a valuable proof-of-concept of robot-robustness and highlights the importance of passive compliance in small quadrupedal robots. This passive adaptability is providing an important fail-safe if more sophisticated control (to be implemented in the future) might fail. Both motions, **sitting/lying down and standing up**, were achieved in ideal conditions, on flat ground without any inclinations of the robot body or even lying on the back. The robot was able to repeatedly move from one posture to the other and start trotting afterward. To further enhance the motion-sequences, especially when not in an ideal position (e.g., on the side), further sensorization with an IMU as well as an active spine are needed.

Conclusively, Serval has the potential for agile locomotion by showing versatility within a basic setup. Relatively stiff legs allow for good shock absorption in a higher weight-range and possibly fast extension (explosive behavior) in low load cases. Modular design and adaptability of spring-stiffnesses enable experimental tuning for our agility tasks efficiently. Turning and locomotion in difficult terrain are in principle possible via different strategies, leveraging the high ROM in AA, spine, and legs. To enhance overall performance and protect the robot from failure, reflexive mechanisms as in Oncilla, based on appropriate sensory feedback (e.g. GRF and body inertia/heading), should extend the robot control. Additionally, to complete the level of mechatronic complexity needed for versatility and agility, we believe that freeing the active compliant spine and integrating an appropriate controller is advantageous. Investigating anisotropic friction, geometry, and stiffness of the feet to allow defined and efficient propulsion should be addressed as well.

Acknowledgements. We thank the "Bewegungslabor (OpenLab) der westfälischen Wilhelms Universität Münster", especially Dr. Marc de Lussanet and Prof. Dr. Heiko Wagner. We wish to thank the owners of the dogs for providing the experimental subjects. This collaborative work was financially supported by the NCCR Robotics and gkf Gesellschaft für kynologische Forschung. We thank the editors and reviewers for their constructive criticism.

References

1. Eckert, P., Ijspeert, A.J.: Benchmarking agility for multi-legged terrestrial robots. IEEE Trans. Robot. 8 (2018, in progress)
2. Ajallooeian, M.: Pattern generation for rough terrain locomotion with quadrupedal robots. Ph.D. thesis, EPFL (2015)
3. Full, R.J., Koditschek, D.E., Full, R.J.: Templates and anchors: neuromechanical hypotheses of legged locomotion on land. J. Exp. Biol. **2**(12), 3–125 (1999)
4. Ijspeert, A.J.: Biorobotics: using robots to emulate and investigate agile locomotion. Science **346**(6206), 196–203 (2014)
5. Seok, S., Wang, A., Chuah, M.Y., Otten, D., Lang, J., Kim, S.: Design principles for highly efficient quadrupeds and implementation on the MIT Cheetah robot. In: Proceedings of IEEE International Conference on Robotics and Automation, pp. 3307–3312. IEEE, May 2013
6. Park, H.W., Park, S., Kim, S.: Variable-speed quadrupedal bounding using impulse planning: untethered high-speed 3D running of MIT Cheetah 2. In: Proceedings of IEEE International Conference on Robotics and Automation, pp. 5163–5170, May–June 2015
7. Hutter, M.: ANYmal - A Highly Mobile and Dynamic Quadrupedal Robot. Arbeitsberichte Verkehrs- und Raumplanung, IVT, ETH Zurich, vol. 544, pp. 1–25 (2009)
8. Hutter, M., Gehring, C., Höpflinger, M.A., Blösch, M., Siegwart, R.: Toward combining speed, efficiency, versatility, and robustness in an autonomous quadruped. IEEE Trans. Robot. **30**(6), 1427–1440 (2014)

9. Poulakakis, I., Smith, J.A., Buehler, M.: Modeling and experiments of untethered quadrupedal running with a bounding gait: the scout II robot. Int. J. Robot. Res. **24**(4), 239–256 (2005)

10. Pusey, J.L., Duperret, J.M., Haynes, G.C., Knopf, R., Koditschek, D.E.: Free-standing leaping experiments with a power-autonomous elastic-spined quadruped. In: SPIE Defense, Security, and Sensing, vol. 8741, p. 87410W (2013)

11. Kimura, H., Fukuoka, Y., Cohen, A.H.: Adaptive dynamic walking of a quadruped robot on natural ground based on biological concepts. Int. J. Robot. Res. **26**(5), 475–490 (2007)

12. Fukuoka, Y., Kimura, H.: Dynamic locomotion of a biomorphic quadruped Tekken robot using various gaits: walk, trot, free-gait and bound. Appl. Bionics Biomech. **6**(1), 63–71 (2009)

13. Iida, F., Pfeifer, R.: Cheap rapid locomotion of a quadruped robot: self-stabilization of bounding gait. In: Proceedings of the 8th International Conference on Intelligent Autonomous Systems (IAS-8), vol. 8, pp. 642–649 (2004)

14. Iida, F., Gómez, G., Pfeifer, R.: Exploiting body dynamics for controlling a running quadruped robot. In: Proceedings of 2005 International Conference on Advanced Robotics, ICAR 2005, vol. 2005, pp. 229–235. IEEE (2005)

15. Spröwitz, A., Tuleu, A., Vespignani, M., Ajallooeian, M., Badri, E., Ijspeert, A.J.: Towards dynamic trot gait locomotion: design, control, and experiments with Cheetah-cub, a compliant quadruped robot. Int. J. Robot. Res. **32**(8), 932–950 (2013)

16. Khoramshahi, M., Sprowitz, A., Tuleu, A., Ahmadabadi, M.N., Ijspeert, A.J.: Benefits of an active spine supported bounding locomotion with a small compliant quadruped robot. In: Proceedings of IEEE International Conference on Robotics and Automation, pp. 3329–3334 (2013)

17. Sprowitz, A.T., et al.: Oncilla robot: a versatile open-source quadruped research robot with compliant pantograph legs. Front. Robot. AI **5**, 67 (2018)

18. Eckert, P., Sprowitz, A., Witte, H., Ijspeert, A.J.: Comparing the effect of different spine and leg designs for a small bounding quadruped robot. In: Proceedings of IEEE International Conference on Robotics and Automation, vol. 2015, pp. 3128–3133, June 2015

19. Weinmeister, K., Eckert, P., Witte, H., Ijspeert, A.J.: Cheetah-cub-S: steering of a quadruped robot using trunk motion. In: 2015 IEEE International Symposium on Safety, Security, and Rescue Robotics (SSRR), pp. 1–6 (2015)

20. Witte, H., et al.: Transfer of biological principles into the construction of quadruped walking machines. In: Proceedings of the 2nd International Workshop on Robot Motion and Control, RoMoCo 2001, pp. 245–249 (2001)

21. Söhnel, K., Andrada, E., De Lussanet, M.H.E., Wagner, H., Fischer, M.S.: Kinetics in Jumping Regarding Agility Dogs (2017)

22. Tuleu, A.: Hardware, software and control design considerations towards low-cost compliant quadruped robots. Ph.D. thesis, EPFL (2016)

23. Horvat, T., Karakasiliotis, K., Melo, K., Fleury, L., Thandiackal, R., Ijspeert, A.J.: Inverse kinematics and reflex based controller for body-limb coordination of a salamander-like robot walking on uneven terrain. In: IEEE International Conference on Intelligent Robots and Systems, vol. 2015, pp. 195–201. IEEE, September–December 2015

24. Smith, J.L., Carlson-Kuhta, P., Trank, T.V.: Forms of forward quadrupedal locomotion. III. A comparison of posture, hindlimb kinematics, and motor patterns for downslope and level walking. J. Neurophysiol. **79**(4), 1702–1716 (1998)

25. Schmerbauch, A.E.M., Eckert, P., Witte, H., Ijspeert, A.J.: Implementation and analysis of rich locomotion behavior on the bio-inspired, quadruped robot Serval (2017)

Minimal Model for Body–Limb Coordination in Quadruped High-Speed Running

Akira Fukuhara[✉], Yukihiro Koizumi, Shura Suzuki, Takeshi Kano, and Akio Ishiguro

Research Institute of Electrical Communication, Tohoku University, 2-1-1 Katahira, Aoba-ku, Sendai 980-8577, Japan
a.fukuhara@riec.tohoku.ac.jp
http://www.cmplx.riec.tohoku.ac.jp

Abstract. Cursorial quadrupeds exploit their limbs and bodies (i.e., body–limb coordination) to achieve faster locomotion speed when compared to that with only limbs. Extant studies examined various legged robots that utilize flexible spine bending. However, the control principle of body–limb coordination is not established to date. This study proposes a novel control scheme for body–limb coordination in which all degrees of freedom of the entire body aid each other in achieving higher performance. The 2D simulation results indicate that mutual sensory feedback between the limb and spine plays essential roles in generating their adaptive locomotion patterns in response to physical situations of the body parts and thereby in achieving faster locomotion speeds.

Keywords: Quadruped locomotion · Body–limb coordination
High-speed locomotion

1 Introduction

Running faster is a critical ability for quadrupeds to survive in the wild. For example, cheetahs and dogs utilize limbs and also actively bend and extend their flexible spine in the pitching direction to increase stride length when compared to that with only limbs [1–3]. Additionally, even small mammals exhibit coordination patterns between the limb and body, i.e., body–limb coordination [4,5]. The motivation of the present study involves understanding the body–limb coordination mechanism underlying quadrupedal locomotion and establishing a design principle for robots that coordinate whole bodily degrees of freedom to achieve faster locomotion speed.

Inspired by the high-speed quadrupedal running, several roboticists developed legged robots with an actuated spine joint. A passive or open-loop controlled spine joint aids robots in running faster under certain conditions (for e.g., locomotion frequency and stiffness of the spine joint) [6–13]. Furthermore,

© Springer Nature Switzerland AG 2018
P. Manoonpong et al. (Eds.): SAB 2018, LNAI 10994, pp. 56–65, 2018.
https://doi.org/10.1007/978-3-319-97628-0_5

a closed-loop control mechanism was proposed in which the coordination patterns of the limb and spine are generated by the reflex chains [14]. However, the body–limb coordination of the robots is still less adaptive and versatile when compared with the locomotion of the cursorial quadrupeds. A challenge relates to the design of sensory feedback mechanism to generate adaptive body–limb coordination based on the physical situation of a robot because the motion of the limb and spine significantly influence each other.

With respect to the aforementioned issue, we propose a novel concept of body–limb coordination mechanism in which all degrees of freedom of the entire body aid each other in enhancing locomotion ability [15,16]. The key feature of the proposed model involves mutual sensory feedback between the limb and body. The mutual feedback allows each body part to modulate its motion based on the physical situation of other parts. In the previous study [15,16], salamander-like robots combined legged locomotion and undulatory locomotion (spine bending motion in the yaw axis) depending on their physical situations, resulting in reproduction of primitive tetrapod walking, e.g., salamander and polypterus. In this study, we design a minimal body–limb coordination mechanism for mammals' faster running and evaluate it by performing a 2D simulation. The results suggest that bilateral sensory feedback terms between the limbs and the spine are essential in achieving the fastest locomotion speed.

2   Model

2.1   Mechanical Model

The body structure of a robot is simplified as shown in Fig. 1 to extract minimal control mechanism for body–limb coordination. The robot consists of simple fore and hind limbs and an actuated trunk and is modeled as a 2D mass-spring-damper system. Point-masses are located at the shoulder, hip, spine, fore foot, and hind foot. The point-masses are connected with a spring and damper. The limb includes two degrees of freedoms (DOFs) as follows: the shoulder (hip) joint consists of a rotary spring and damper while the limb extends and contracts via a prismatic spring and damper. Furthermore, the trunk includes one DOF as follows: the spine joint consists of a rotary spring and damper. The prismatic spring and damper between the spine point-mass and shoulder (hip) are set as stiff to ensure that the link of the trunk is rigid.

By changing the natural angle and length of the rotary and prismatic spring, the position of the foot traces a specific trajectory as follows:

$$\bar{\theta}_i = \theta_i^{os} + C^{rot} \cos \phi_i, \tag{1}$$

$$\bar{l}_i = \begin{cases} l^{os} - C_{sw}^{pri} \sin \phi_i \; (\sin \phi_i > 0), \\ l^{os} - C_{st}^{pri} \sin \phi_i \; (\sin \phi_i < 0), \end{cases} \tag{2}$$

where i denotes the index of the limbs ($i = fore, hind$), $\bar{\theta}_i$ denotes the target angle of the shoulder (hip) joint, θ_i^{os} denotes a constant value for angular offset of the limb joint, C^{rot} denotes a constant value to define the amplitude of rotary motion,

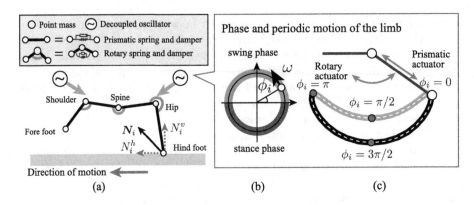

Fig. 1. Mechanical model of the robot. (a) Mass-spring-damper system. (b) Phase oscillator. (c) Foot trajectory.

ϕ_i denotes a phase of oscillator of the limb, l_{os} denotes a constant value for the offset of the prismatic joint, and C_{sw}^{pri} (C_{st}^{pri}) denotes a constant value to define the amplitude of the prismatic motion during contraction (extension). When $\sin \phi_i > 0$, the length of the limb decreases and it tends to lift off the ground (swing phase). Conversely, when $\sin \phi_i < 0$ the limb increases in length and tends to remain on the ground to support the body (stance phase). To ensure the ground clearance during the swing phase, C_{sw}^{pri} is designed to be larger than C_{st}^{pri}.

The torque, τ_i, and force, F_i, of the limb is calculated as follows:

$$\tau_i = K^{rot}(\bar{\theta}_i - \theta_i) - D^{rot}\dot{\theta}_i, \tag{3}$$
$$F_i = K^{pri}(\bar{l}_i - l_i) - D^{pri}\dot{l}_i, \tag{4}$$

where K^{rot} and D^{rot} denote spring and damper coefficients of hip (shoulder) joint of the limb, K^{pri} and D^{pri} denote spring and damper coefficients of prismatic joint of the limb, θ_i denotes the actual angle of the hip (shoulder) joint, and l_i denotes the actual length of the limb, respectively.

2.2 Control Model

The purpose of this study involves designing a minimal control mechanism that enables the robot to generate adaptive and coordinated locomotion patterns between the limbs and spine based on their physical situations. In our model, sensory feedback plays essential roles in adaptive locomotion patterns as shown in Fig. 2. Specifically, the motions of the limb and trunk are modulated by sensory information from self and other body parts. The details of feedback rules are discussed in the following subsections.

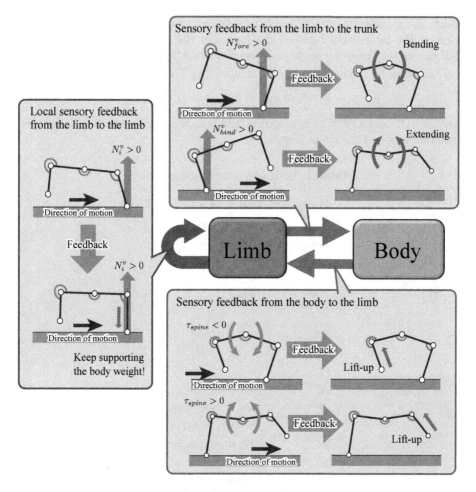

Fig. 2. Overview of the proposed body–limb coordination mechanism

Body Control. Based on the behaviors, we hypothesize that sensory feedback from limbs to the body should be designed such that spine motion increases stride length. The reference angle of the spine is described as follows:

$$\bar{\theta}_{spine} = \theta_{spine}^{os} + \sigma_1(N_{hind}^v - N_{fore}^v), \qquad (5)$$

where $\bar{\theta}_{spine}$ denotes the natural angle of the rotational spring at the spine point-mass, θ_{spine}^{os} denotes the offset angle, σ_1 denotes the constant weighting value for sensory feedback, and N_{fore}^v and N_{hind}^v represents the vertical component of ground reaction force (GRF) applied at the forelimb and hindlimb, respectively. The effects of the sensory feedback terms are explained as follows. When the forelimb experiences GRF ($N_{fore} > 0$), the spine joint bends and the lower body is pulled forward. Conversely, when the hindlimb experiences GRF ($N_{hind} > 0$), the spine joint extends and the upper body is pushed forward. The feedback term reflects the magnitude of GRF, and thus, the motion of the spine is modulated based on the interaction between the limb and the ground, e.g., a decrease in GRF causes the reference angle to correspond to θ_{spine}^{os}.

The torque of the active spine is calculated as follows:

$$\tau_{spine} = K^{spine}(\bar{\theta}_{spine} - \theta_{spine}) - D^{spine}\dot{\theta}_{spine}, \tag{6}$$

where K^{spine} and D^{spine} denote spring and damper coefficients, θ_{spine} denotes the actual angle of the spine joint.

Table 1. Body and control parameters.

Parameter	Value	Unit
Total mass	53.0	[kg]
Body length	1.1	[m]
l_{os}	0.5	[m]
C_{sw}^{pri}	0.2	[m]
C_{st}^{pri}	0.065	[m]
C^{rot}	1.0	[rad]
θ_{fore}^{os}	$\pi/2$	[rad]
θ_{hind}^{os}	$3\pi/2$	[rad]
θ_{spine}^{os}	π	[rad]
K^{pri}	4000.0	[N/m]
D^{pri}	400.0	[Ns/m]
K^{rot}	3600.0	[Nm/rad]
D^{rot}	230.0	[Nms/rad]
K^{spine}	8000.0	[Nm/rad]
D^{spine}	400.0	[Nms/rad]
ω	25.0	[rad/s]
σ_2	0.025	[rad/Ns]

Limb Control. Quadrupeds exhibit versatile coordination pattern between the limbs (interlimb coordination) based on locomotion speed, species, and environment. The adaptive locomotion pattern patterns are partially generated by the decentralized control system, e.g., central pattern generators, CPGs, and local reflex i.e., sensory feedback [17]. In a previous study, we proposed a simple interlimb coordination mechanism that enabled a robot to exhibit spontaneous gait transition with increases in locomotion speed [18,19]. Based on the previous interlimb coordination mechanism, the time evaluation of the phase of the limb controller, ϕ_i, is described as follows:

$$\dot{\phi}_{fore} = \omega - \sigma_2 N_{fore}^v \cos\phi_{fore} + \sigma_3 \max[0, \tau_{spine}] \cos\phi_{fore}, \tag{7}$$

$$\dot{\phi}_{hind} = \omega - \sigma_2 N_{hind}^v \cos\phi_{hind} - \sigma_3 \min[0, \tau_{spine}] \cos\phi_{hind}, \qquad (8)$$

where ω denotes an intrinsic angular velocity that is related to the locomotion frequency, σ_2 and σ_3 denote constant values for sensory feedback terms based on sensory information from own limb and the trunk. The effects of feedback terms are explained in the following section.

When σ_3 is zero, Eqs. (7) and (8) are essentially identical to the previous interlimb coordination model. In the previous model [18], the motion of each limb is modulated based on the magnitude of GRF, N_i, such that the loaded limb tends to keep supporting the body weight. Specifically, while $N_i > 0$, the second term on the right side of Eqs. (7) and (8) increases ϕ_i during the second and third quadrants, and it decreases ϕ_i during the first and fourth quadrants.

In order to generate the adaptive body–limb coordination, the third term of Eqs. (7) and (8) reflects the situation of the spine. We assumed that the limbs should stay steady during swing phase when the spine actively bends/extends. Specifically, when the spine tends to extend ($\tau_{spine} > 0$), the forelimb waits around the middle of the swing phase. Conversely, when the spine tends to bend ($\tau_{spine} < 0$), the hindlimb waits around the middle of the swing phase. The coordination patterns for faster locomotion are generated based on the situations via sensory feedback between the limbs and the trunk as shown in Eqs. (5)–(8).

3   Simulation

In order to verify the locomotion patterns via the proposed model, we conducted a 2D simulation with mechanical and control parameters as shown in Table 1. The mass and size of the robot are set based on cheetahs [20]. The control parameters are set through try-and-error. In order to investigate the effects of feedback terms between the limb and body, the robot runs only via local sensory feedback from limb to limb, i.e., $(\sigma_1, \sigma_2, \sigma_3) = (0.0, 2.50 \times 10^{-2}, 0.0)$. Subsequently, weighting values for body–limb coordination are increased as follows: σ_1 changes from 0.0 to 1.60×10^{-4} [rad/N] and σ_3 changes from 0.0 to 1.15×10^{-1} [rad/Nms]. It should be noted that input ω is constant during the simulation ($\omega = 25.0$ [rad/s]).

Figure 3 shows the changes in the locomotion speed during the simulation. When the spine operated as a passive joint ($\sigma_1 = 0.0$), the average locomotion speed corresponded to 9.58 [leg length/s] (leg length is assumed as $l^{os} = 0.5$ [m]). When the spine operated as an actuated joint ($\sigma_1 = 1.60 \times 10^{-2}$), the locomotion speed increased from 9.58 to 10.28 [leg length/s]. Furthermore, when the feedback from the spine to the limb was performed, the robot exhibited the fastest locomotion speed (11.98 [leg length/s]). Figure 4 shows the stick diagrams of the robot with each set of control parameters. The distance of the footprint of a limb, i.e., stride length, increases with the application of the additional feedback mechanisms between the limb and body. The locomotion patterns of the robot are shown in a video clip (https://fsa.fir.riec.tohoku.ac.jp/fircloud/index.php/s/zrtH04rmV1D6VLh).

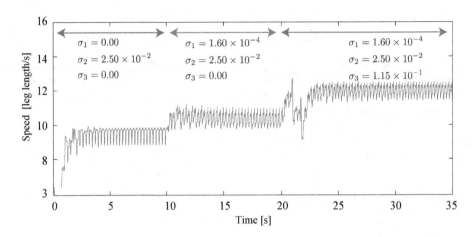

Fig. 3. Locomotion speed and locomotion patterns during simulation. The locomotion speed is normalized with leg length ($l^{os} = 0.5\,[\text{m}]$).

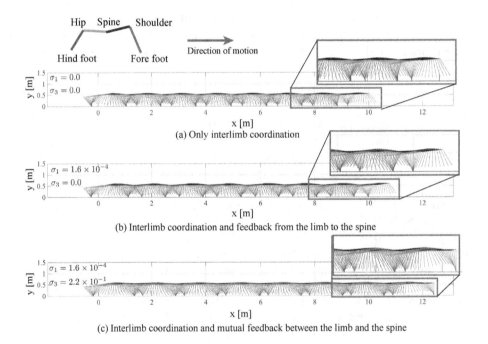

Fig. 4. Locomotion patterns with each set of parameter gains.

Figure 5 shows the details of the body limb coordination of the robot. Without the bilateral feedback mechanism between the limb and spine (Fig. 5(a)), the spine essentially maintained a straight posture. In contrast, the feedback from the limb to the spine enabled the robot to actively bend and extend the spine in

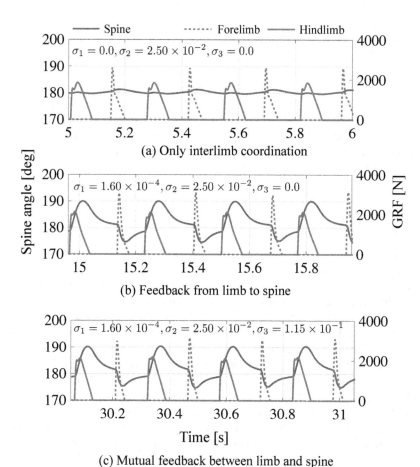

Fig. 5. Spine angle during locomotion with each set of parameter gains.

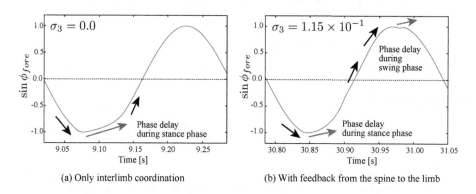

Fig. 6. Comparison between phase modulation between (a) only interlimb coordination and (b) with feedback from the spine to the limb.

response to GRF applied by each limb, thereby resulting in increases in locomotion speed (Fig. 5(b) and (c)). These results suggest that the sensory feedback from the limb to spine play essential role for actuating the trunk appropriately.

Furthermore, the sensory feedback from the trunk to limb is essential to increase locomotion speed. Figure 6 shows changes of the phase of the controller without and with sensory feedback from the spine to the limb. In both cases, the phase of the controller, ϕ_i, is delayed during the stance phase by the local sensory feedback in the limb (Fig. 6(a)). Additionally, the feedback from the spine to the limb delays the phase during swing phase (Fig. 6(b)). The phase delay in the swing phase enables the limb to stay in the air, and thus increases the stride length.

4   Conclusion

The goal of the study involves understanding the body–limb coordination mechanism underlying quadrupedal locomotion for realizing faster running robots. For this purpose, we proposed a minimal body–limb coordination mechanism in which different types of locomotor organs, namely the limb and spine, aid each other in increasing the stride length. The results of a 2D simulation indicated that the robot achieved faster running via mutual sensory feedback between the limb and spine. Addition to reproduction of mammals' faster running, the concept of body-limb coordination can reproduce amphibians' walking, e.g., combination of legged and undulatory locomotion (spine bending in yaw axis) [15,16]. Because adaptive locomotions of different species, e.g., mammals' running and amphibians' walking, are well reproduced via the mutual sensory feedback between the limbs and the body, we believe that our concept could help to understand a common principle how animals recruit their bodily degrees of freedom to achieve more adaptive locomotion.

A future study will involve applying the proposed model to a four-legged robot and investigating the effects of body–limb coordination on high-speed gaits as to whether body–limb coordination is a crucial factor in understanding the mechanism of transverse gallop and rotary gallop.

Acknowledgments. This work was supported by Japan Science and Technology Agency, CREST (JPMJCR14D5).

References

1. Bertram, J.E., Gutmann, A.: Motions of the running horse and cheetah revisited: fundamental mechanics of the transverse and rotary gallop. J. Roy. Soc. **6**, 549–559 (2008)
2. Maes, L.D., Herbin, M., Hackert, R., Bels, V.L., Abourachid, A.: Steady locomotion in dogs: temporal and associated spatial coordination patterns and the effect of speed. J. Exp. Biol. **211**, 138–149 (2008)
3. Biancardi, C.M., Minetti, A.E.: Biomechanical determinants of transverse and rotary gallop in cursorial mammals. J. Exp. Biol. **215**, 4144–4156 (2012)

4. Schiling, N., Hackert, R.: Sagittal spine movements of small therian mammals during asymmetrical gaits. J. Exp. Biol. **209**, 3925–3939 (2006)
5. Schilling, N., Carrier, D.R.: Function of the epaxial muscles in walking, trotting and galloping dogs: implications for the evolution of epaxial muscle function in tetrapods. J. Exp. Biol. **213**, 1490–1502 (2010)
6. Deng, Q., Wang, S., Xu, W., Mo, J., Liang, Q.: Quasi passive bounding of a quadruped model with articulated spine. Mech. Mach. Theor. **52**, 232–242 (2012)
7. Cao, Q., Poulakakis, I.: Quadrupedal running with a flexible torso: control and speed transitions with sums-of-squares verification. Artif. Life Robot. **21**, 384–392 (2016)
8. Seok, S., Wang, A., Chuah, M.Y., Otten, D., Lang, J., Kim, S.: Design principles for highly efficient quadrupeds and implementation on the MIT cheetah robot. In: Proceedings of IEEE ICRA 2013, pp. 3307–3312 (2013)
9. Eckert, P., Spröwitz, A., Witte, H., Ijspeert, A.J.: Comparing the effect of different spine and leg designs for a small bounding quadruped robot. In: Proceedings of IEEE ICRA 2015, pp. 3128–3133 (2015)
10. Pouya, S., Khodabakhsh, M., Spröwitz, A., Ijspeert, A.J.: Spinal joint compliance and actuation in a simulated bounding quadruped robot. Auton. Robot. **41**, 437–452 (2017)
11. Wang, C., Zhang, T., Wei, X., Long, Y., Wang, S.: Dynamic characteristics and stability criterion of rotary galloping gait with an articulated passive spine joint. Ad. Robot. **31**(4), 168–183 (2017)
12. Duperret, J., Koditschek, D.E.: Empirical validation of a spined sagittal-plane quadrupedal model. In: Proceedings of IEEE ICRA 2017, pp. 1058–1064 (2017)
13. Chen, D., Li, N., Wang, H., Chen, L.: Effect of flexible spine motion on energy efficiency in quadruped running. J. Bionic Eng. **14**, 716–725 (2017)
14. Wang, C., Wang, S.: Bionic control of cheetah bounding with a segmented spine. Appl. Bionics Biomech. **2016**, 1–12 (2016). Article No. 5041586
15. Shura, S., Fukuhara, A., Owaki, D., Kano, T., Ijspeert, A.J., Ishiguro, A.: A minimal model for body-limb coordination in quadruped locomotion. In: The 8th International Symposium on Adaptive Motion of Animals and Machines, p. 39 (2017)
16. Shura, S., Fukuhara, A., Owaki, D., Kano, T., Ijspeert, A.J., Ishiguro, A.: A simple body-limb coordination model that mimics primitive tetrapod walking. In: Proceedings of the SICE Annual Conference 2017, pp. 12–14 (2017)
17. Grillner, S.: Locomotion in vertebrates: central mechanisms and reflex interaction. Physiol. Rev. **55**(2), 247–304 (1975)
18. Owaki, D., Kano, T., Nagasawa, K., Tero, A., Ishiguro, A.: Simple robot suggests physical interlimb communication is essential for quadruped walking. J. Roy. Soc. Interface **10**(78), 20120669 (2013)
19. Owaki, D., Ishiguro, A.: A quadruped robot exhibiting spontaneous gait transitions from walking to trotting to galloping. Sci. Rep. **7**, 77 (2017). https://doi.org/10.1038/s41598-017-00348-9
20. Hydson, P.S., Corr, S.A., Davis, R.C., Clancy, S.N., Lane, E., Wilson, A.M.: Functional anatomy of the cheetah (Acinonyx jubatus) forelimb. J. Anat. **218**(4), 375–385 (2011)

An Active Efficient Coding Model of Binocular Vision Development Under Normal and Abnormal Rearing Conditions

Lukas Klimmasch[1(✉)], Johann Schneider[1], Alexander Lelais[1], Bertram E. Shi[2], and Jochen Triesch[1]

[1] Frankfurt Institute for Advanced Studies, Frankfurt am Main, Germany
`klimmasch@fias.uni-frankfurt.de`
[2] Hong Kong University of Science and Technology, Hong Kong, China

Abstract. The development of binocular vision encompasses the formation of binocular receptive fields tuned to different disparities and the calibration of accurate vergence eye movements. Experiments have shown that this development is impaired when the animal is exposed to certain abnormal rearing conditions such as growing up in an environment that is deprived of horizontal or vertical edges. Here we test the effect of abnormal rearing conditions on a recently proposed computational model of binocular development. The model is formulated in the Active Efficient Coding framework, a generalization of classic efficient coding ideas to active perception. We show that abnormal rearing conditions lead to differences in the model's development that qualitatively match those seen in animal experiments. Furthermore, the model predicts systematic changes in vergence accuracy due to abnormal rearing. We discuss implications of the model for the treatment of developmental disorders of binocular vision such as amblyopia and strabismus.

Keywords: Receptive field development · Sparse coding
Abnormal rearing condition · Active efficient coding · Stereopsis
Vergence

1 Introduction

The efficient coding hypothesis states that biological sensory systems represent sensory information efficiently by using neural codes that exploit redundancies in the statistics of the sensory signals [1,2]. Perception is an active process, however, and the statistics of sensory signals are also a function of the organism's behavior. In primate vision, for example, the statistics of binocular disparities depend on both the distances of objects in the environment and on the individual's vergence eye movements.

ⓒ Springer Nature Switzerland AG 2018
P. Manoonpong et al. (Eds.): SAB 2018, LNAI 10994, pp. 66–77, 2018.
https://doi.org/10.1007/978-3-319-97628-0_6

Active Efficient Coding (AEC) is a recent generalization of the classic efficient coding hypothesis to active perception [3,4]. It states that biological sensory systems jointly optimize their sensory representation and their behavior to achieve an efficient encoding of their sensory environment. This is realized by learning an efficient sparse code for sensory signals while learning to move the sense organs such that the sparse coding model works efficiently. Along these lines fully self-calibrating models for the development of active binocular vision [3,5], active motion vision (motion tuning, pursuit, optokinetic reflex) [4,6], accommodation control [7], torsional eye movements (in preparation) and combinations thereof [8] have been proposed. Such models have explained certain aspects of the development of binocular vision in mammals [9] and they have been validated on physical robots [4,10].

Interestingly, the development of biological vision systems can be derailed by abnormal visual input during the so-called critical period of visual development. A range of experiments have established how systematic alterations to an animal's visual input or complete deprivation from visual input harm the developing visual representations and behavior [11–15]. An important benchmark for computational models of visual development is therefore whether altered rearing conditions affect the model's development in a similar way to what is seen in animal experiments [16].

Here we test the effects of a range of altered rearing conditions on a recently proposed AEC model for the development of stereopsis through the simultaneous learning of disparity tuning and vergence eye movements [17]. We find that altered rearing conditions systematically change both the model's receptive field properties and its ability to perform accurate vergence eye movements. Importantly, the effects of the altered rearing conditions qualitatively match those observed in animal experiments. This finding lends additional support to AEC as a theoretical framework for understanding the development of active perception.

2 Methods

2.1 Model and Learning Environment

The model and the environment have been described in detail in [17]. Here, we provide a brief overview of the model and focus on the simulation of the altered rearing conditions.

The model comprises an agent in a simulated 3-D environment (Fig. 1A). It is implemented using OpenEyeSim [18], a simulation software combining a detailed biomechanical model of the human extra-ocular eye muscles and the rendering of a 3-D environment. During learning, textured planar objects are presented to the agent at different distances. We use 100 randomly chosen images from the McGill database [19] as textures. Time runs in discrete time steps corresponding to roughly 30 ms. An object is present for 10 time steps, which we refer to as one fixation, after which it is replaced by a different object at a random distance chosen uniformly between 0.5 m and 6.0 m. In each iteration of one fixation,

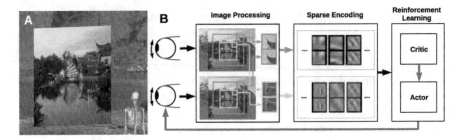

Fig. 1. Model overview. A The agent looking at a planar object in the OpenEyeSim environment. **B** Processing of binocular input by the agent. Binocular patches are extracted at a coarse scale (green boxes) and a fine scale (yellow boxes) with different resolutions. These patches are encoded by activations of basis functions via sparse coding and combined with the muscle activations to generate a state vector (black arrow on the right). This vector is fed into an actor-critic reinforcement learning model which learns to perform vergence eye movements (red arrow). The critic evaluates the actor's actions by calculating the temporal difference error (blue arrow). The negative reconstruction error of the sparse coding model is a measure of the system's coding efficiency and is used as a reward signal. Adapted from [17]. (Color figure online)

the agent acquires and processes images from its two eyes, encodes them and executes a vergence eye movement command, after which the next iteration starts. A schematic of this process is depicted in Fig. 1B.

The binocular images are dissected into two sets of $2 \times 8 \times 8$ binocular pixel patches: one *coarse scale* set, where the edge of a patch covers a visual angle of 7.0°, and a *fine scale* set, where the edge of a patch covers 1.76°. Patches are normalized to zero mean and unit variance. Using different scales improves robustness of the vergence learning [10] and allows us to model the development of receptive fields (RFs) at different spatial resolutions.

Each of the two sets of binocular patches is encoded using the matching pursuit algorithm [20] that aims to reconstruct the original input patch as a sparse combination of $N_{\text{active}} = 10$ basis functions (BFs) chosen from a learned over-complete dictionary of $N_B = 400$ BFs. These correspond to binocular simple cell RFs in primary visual cortex and will be analyzed in detail below. In the beginning of training, analogous to the state just before eye opening, we initialize the BFs with random Gabor functions. Specifically, both the left eye and the right eye component of a binocular basis function have the shape of a Gabor function, but the two Gabor functions have independently drawn random orientations. We have verified that the results below can also be achieved when BFs are initialized as Gaussian white noise. The use of random Gabors makes the vergence learning more stable and is biologically more plausible [21].

The matching pursuit sparse coding algorithm computes two vectors (one for the coarse and one for the fine scale) that contain the activations of all N_B BFs averaged over the whole image. These vectors are squared and concatenated and serve — together with the current activation of the eye muscles — as the state representation. This representation is fed into a reinforcement learning (RL) algorithm [22,23], which selects changes in muscle activation in a continuous action space. We use an actor-critic RL architecture, namely the CACLA+VAR algorithm [24]. Actor and critic are function approximators, represented by two feed-forward neural networks with three and two layers, respectively. Explorative Gaussian noise is added to the actor's output to enable the discovery of better actions. The critic estimates the value function and instructs the actor via the temporal difference (TD) error. The special property of this algorithm is that the actor is updated only after a positive TD error. We use the parameters specified in [17].

The reinforcement signal used by the model is the negative reconstruction error that is calculated as the difference between the original input images and the reconstruction by the sparse coding algorithm. Since the total amount of information in the extracted image patches is approximately constant and the sparse coding model always uses a fixed number of N_{active} basis functions to encode the patches, the negative reconstruction error of the sparse coding model is effectively a measure of the system's coding efficiency.

While the generated movements are random in the beginning, the learning agent soon discovers that aligning the eyes on the stimulus plane will improve the neural reconstruction and therefore its coding efficiency. Thus the agent will learn to reliably generate appropriate vergence eye movements [3,10,17]. As vergence behavior improves, the statistics of observed disparities change as well. The orientations of right eye and left eye BFs align [25] and more and more BFs become tuned to small horizontal disparities. This improved representation of small disparities allows for even better vergence control and so on. We show the model's improving capacity for accurate vergence control by plotting the vergence error, the difference between the correct vergence angle (which is a function of the object's distance and the distance between the two eyes) and the actual vergence angle at the end of a fixation. Each learning experiment lasts 500 000 iterations (50 000 fixations). With humans taking 0.3 s per fixation on average, this would correspond to a total learning time of 4.2 h— well below the ~3 months that is available to human infants for learning vergence control.

At the end of the training procedure the model is tested on a different set of 40 images from the McGill Database [19]. The execution of explorative actions in the actor is disabled during these tests, i.e., we consider the "greedy" policy. Each of the testing stimuli is positioned at $[0.5, 1, \ldots, 6]$ m and the eyes are initialized with 7 different vergence errors from $[-2, \ldots, 2]$ deg. We record the vergence error after one fixation (10 iterations). To compare the normally reared model's performance at this procedure with that of the different rearing conditions, we use the student's t-test.

2.2   Simulation of Different Rearing Conditions

In biological experiments, different rearing conditions are frequently implemented by fixing cylindrical lenses in front of the animal's eyes [12,13]. Such lenses strongly blur edges of a particular orientation. We simulate such rearing conditions by convolving the input images with elongated Gaussian kernels defined by:

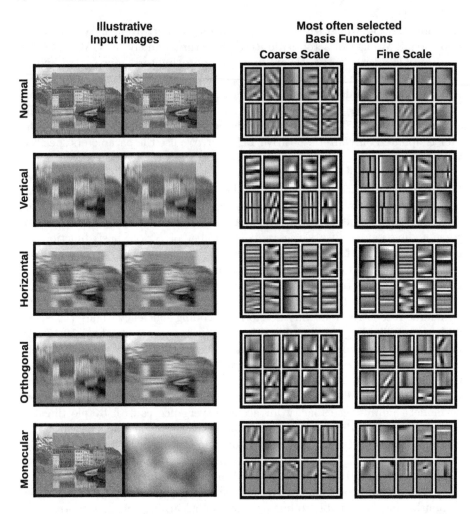

Fig. 2. Input scenarios and learned receptive fields. Left: Illustration of the input under different rearing conditions. In the orthogonal case left and right eye inputs are filtered with kernels rotated by 90°. In the case of monocular deprivation the right eye is blurred. **Right:** Representative examples of binocular BFs for the fine and coarse scale learned under the different rearing conditions. For each BF the left eye and right eye patch are aligned vertically. In each case, the BFs used most frequently by the sparse coding algorithm are shown.

$$K_{\sigma_x, \sigma_y}(x, y) = \exp\left(-\left(\frac{x^2}{2\sigma_x^2} + \frac{y^2}{2\sigma_y^2}\right)\right), \quad (1)$$

where $\sigma_{x/y}$ represent the standard deviation in the horizontal/vertical direction. For example, a kernel with large σ_x will blur any vertical edges such that the resulting image mostly contains horizontal edges. Therefore, when mostly vertical input images (*vertical case*) should be simulated we set σ_x to 0.1 px and σ_y to 33 px. With these parameters, the longer dimension of the Gaussian kernel covers the extent of one image patch at the coarse scale. To simulate the *normal case* we convolve the images with a Gaussian kernel with $\sigma_x = \sigma_y = 0.1$. For simulating the *monocular deprivation case* as in [16], we use $\sigma_x = \sigma_y = 240$ px, which corresponds to the height of the whole input image, for the right input image only. Patches extracted from that image contain only very low spacial frequencies and most of the structure from the original image is removed. Example images after applying the kernels are shown in Fig. 2.

2.3 Analysis of BF Statistics

We analyze the binocularity of the BFs by calculating the binocularity index similar to [16]:

$$BI = \frac{\|w_r\| - \|w_l\|}{\|w_r\| + \|w_l\|} \qquad (2)$$

where $\|w_{l/r}\|$ corresponds to the left (right) part of one BF and $\|.\|$ is the euclidean norm. The BIs of all BFs are binned into 7 binocularity classes with boundaries at $\{-1, -0.85, -0.5, -0.15, 0.15, 0.5, 0.85, 1\}$ as in [16] so that the scale goes from left-monocular (bin 1) over binocular (bin 4) to right-monocular BFs (bin 7).

To determine the orientations of the basis functions we use MATLAB's implementation of the *trust region reflective algorithm* for non-linear curve fitting to fit the Gabor function defined by:

$$G(x, y, \theta, \lambda, \psi, \sigma, \gamma) = \exp\left(-\frac{x'^2 + \gamma^2 y'^2}{2\sigma^2}\right) \cos\left(2\pi\frac{x'}{\lambda} + \psi\right), \qquad (3)$$

with $x' = x \cos(\theta) + y \sin(\theta)$ and $y' = -x \sin(\theta) + y \cos(\theta)$. Here, λ denotes the wavelength, ψ the phase offset, σ the standard deviation of the Gaussian envelope, γ the spatial aspect ratio and θ the orientation, where $\theta = 0°$ corresponds to a vertically oriented Gabor function. We initialize the parameters randomly 150 times and fit the function each time either to the left or right patch or to both. For the plots in Fig. 3 we took the θ of the best fit into consideration if its norm of residuals was smaller than 0.2 ($> 90\%$ of all BFs). We only show data for the fine scale since differences across rearing conditions are more obvious there compared to the coarse scale.

3 Results

3.1 Normal Rearing

In the case of normal rearing, Gabor-like BFs develop in the sparse coding model (Fig. 2). As can be seen in Fig. 3, they are tuned to all possible orientations with the number of BFs tuned to vertical and horizontal orientations exceeding that

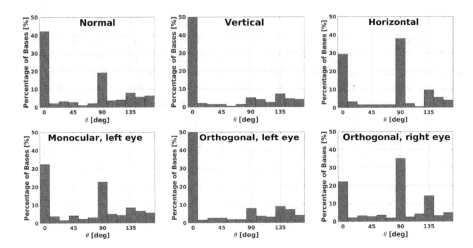

Fig. 3. Orientation distributions for different rearing conditions. The plots are generated by fitting a Gabor function (Eq. 3) to the BFs as described in Sect. 2.3. In the normal, vertical and horizontal case two Gabor functions are fitted to the binocular patches. When the right eye is deprived, the right patches show little structure. In this case the analysis is done only for the left eye patches. In the orthogonal rearing condition the plots for each eye are presented. Here the left eye received only vertical orientations and the right only horizontal ones as illustrated in Fig. 2.

of others. Vertical BFs are most abundant (40%), followed by horizontal ones (25%). An over-representation of horizontal and vertical RFs is observed in the primary visual cortex of cats, humans, monkeys and other species [26] and is known as the *oblique effect*. The large majority of BFs is binocular with a peak at bin 4 (Fig. 4) and no BFs are completely monocular.

As can be seen in Fig. 5, the model learns to correctly verge the eyes onto the stimulus plane. This means that both eyes receive almost identical input at the end of a typical fixation. The high correlation between left and right eye inputs leads to the development of binocular BFs that are correlated in their monocular sub-parts. However, left and right input images are not exactly the same at all times. This is why we also observe BFs that are not perfectly binocular in Fig. 4.

The performance of this model in the testing procedure can be found in Fig. 6. The large majority of vergence errors from all different testing conditions fall below 0.22° which corresponds to a one pixel error in the input image.

3.2 Striped Environment

Vertical Input. In this scenario all horizontal edges are blurred. Figure 3 shows that around 50% of all BFs are tuned to vertical edges and the prominent peak for horizontal edges that is present in the normal case is missing. This demonstrates that the BFs are adapting to the statistics of the visual inputs. The binocularity is very similar to the case with normal input (Fig. 4).

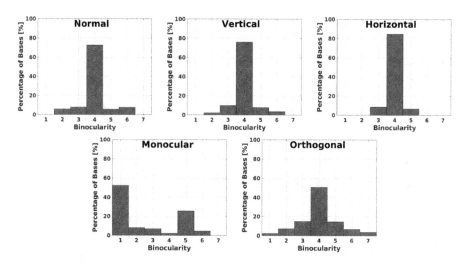

Fig. 4. Histograms of binocularity classes for different rearing conditions. Class 1 corresponds to left monocular BFs, class 4 to binocular BFs, and class 7 to right monocular BFs.

For this rearing condition the model also learns to accurately verge the eyes on the objects as can be seen in Fig. 5. In fact, the model learns slightly faster than in the normal case. Figure 6A shows that the performance in the testing procedure also exceeds that for normal input (student's t test, p-value $< 10^{-6}$).

Horizontal Input. Here, the filtering blurs all vertically oriented edges. The orientation tuning shows that the highest number of orientation-tuned BFs respond to horizontal edges (Fig. 3). But there are still some BFs that are tuned to vertical edges. This is most likely the result of vertical edges being overrepresented in natural images and the standard deviation in the Gaussian kernel being not very big. Nevertheless, there is a clear reduction in the vertically tuned BFs in comparison to the normal input model and a boost in the number of horizontally tuned ones.

Interestingly, the learning of vergence eye movements is still possible in this scenario as shown in Fig. 5 and even seems to be sped up in the beginning. The reasons for this are currently being investigated. We speculate that this might be partly due to a *"less is more"* effect — faster learning for more blurry inputs — as has been observed previously [27]. Additionally, due to down-sampling, the coarse scale is not affected as much as the fine scale from removing the vertical edges and might still be sufficient to guide vergence movements here. Comparing the test results in Fig. 6A with the normal input model, we see that the performance is slightly impaired, but this effect is not significant (p-value = 0.15).

Orthogonal Input. For simulating orthogonal input, we blur horizontal edges in the left eye and vertical edges in the right eye. During training, orthogonal

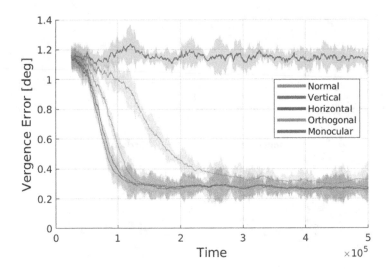

Fig. 5. Vergence performance as a function of training time under different rearing conditions. Depicted are the medians of a moving window of 2500 fixations. Shaded areas correspond to the standard deviation of 5 different simulations.

BFs are acquired (Fig. 2) that have been observed in similar experiments with kittens. [28]. The orientation tuning is estimated separately for the two eyes. As expected, Fig. 3 depicts similar statistics of orientation tuning as for the vertical or horizontal input cases in the corresponding eye. On the other hand, the binocularity is changing noticeably. In Fig. 4 we can observe more monocularly tuned BFs and fewer BFs in class 4 than for the normal input. Furthermore, the model predicts that learning to verge the eyes in this context proceeds more slowly (Fig. 5). At the end of training, the vergence performance is significantly impaired compared to the normal case (Fig. 6A, $p < 10^{-30}$).

3.3 Monocular Deprivation

We simulate different degrees of monocular deprivation by strongly blurring the input to the right eye with different probabilities. Concretely, for every fixation (see Sect. 2.2) the right eye is blurred with a certain probability $\rho_d \in \{0, 0.1, 0.25, 0.5, 0.75, 1\}$. Here, $\rho_d = 1$ corresponds to full deprivation and $\rho_d = 0$ corresponds to no deprivation (analogous to *normal input*).

Interestingly, the orientation tuning seems not to be affected by different degrees of deprivation of one eye. That is why in Fig. 3 we display the results for $\rho_d = 1$ only. The orientation distribution is very similar to the case with unchanged input. Noteworthy is the small decline of the number of BFs tuned to vertical orientations whereas the number of horizontally tuned BFs remains approximately the same. In terms of binocularity (Fig. 4), the model behaves as expected from animal studies: the large majority of BFs fall into class 1, meaning they encode only input from the left eye. Figure 6B shows that the

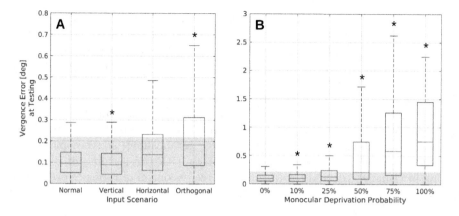

Fig. 6. Absolute vergence error in the testing procedure after training. A For different striped rearing conditions. **B** For different fractions of monocular deprivation. Testing is done without deprivation. Gray bars depict one pixel error in the input image, stars indicate significant deviations from the normal/non-deprived case (p-values < 0.05) and outliers are not depicted.

vergence performance at testing procedure declines with increasing probability of deprivation. A student's t-test revealed significant changes in accuracy (p-values $< 10^{-15}$) in all cases.

4   Discussion

The Active Efficient Coding (AEC) framework posits that biological sensory systems jointly adapt their neural codes and their behavior to optimize their overall coding efficiency. Here we have shown that an AEC model not only describes normal, healthy development of binocular vision but also explains altered developmental trajectories due to abnormal rearing conditions including monocular deprivation or visual environments dominated by a particular orientation. Previous work by Hunt et al. [16] has demonstrated that abnormal rearing conditions can affect the statistics of tuning properties of basis functions learned by different sparse coding models. However, that work did not address the learning of vergence eye movements. In fact, their model implicitly assumed perfect vergence control even under conditions where no disparity tuning develops. In contrast, AEC captures the interdependence of the learning of sensory codes (here: disparity tuning) and behavior (here: vergence eye movements). These findings support AEC as a promising theoretical framework for describing the development of active perception. It will be interesting to see to what extent AEC can also be applied to other sensory modalities.

The results from the analysis of the BFs are in general agreement with those in the work by Hunt et al. [16]. However, the abundance of vertically tuned BFs (Fig. 3) is greater in our model for the normal, horizontal, vertical and orthogonal rearing conditions. This could be due to the different image data bases used in the two studies. Additionally, our results reflect the main finding of similar studies in biological systems [12,29]: The distribution of orientation tuning properties is biased towards the orientations the system was exposed to.

The model presented here has implications for developmental disorders of binocular vision (amblyopia, strabismus) and their treatment. First, it offers an account of the mechanisms through which abnormal visual input impairs the development of disparity tuning and vergence control. Second, the model suggests that any therapy aimed at restoring (or establishing) stereoscopic vision should address (either explicitly or implicitly) both the learning of disparity tuning in visual cortex and the learning of precise control of vergence eye movements.

Acknowledgements. This work was supported by the German Federal Ministry of Education and Research under Grants 01GQ1414 and 01EW1603A, the European Union's Horizon 2020 Grant 713010, the Hong Kong Research Grants Council under Grant 16244416, and the Quandt Foundation.

References

1. Barlow, H.B.: Possible principles underlying the transformations of sensory messages. Sensory Communication (1961)
2. Olshausen, B.A., Field, D.J.: Sparse coding with an overcomplete basis set: A strategy employed by v1? Vision. Res. **37**(23), 3311–3325 (1997)
3. Zhao, Y., Rothkopf, C.A., Triesch, J., Shi, B.E.: A unified model of the joint development of disparity selectivity and vergence control. In: IEEE International Conference on Development and Learning and Epigenetic Robotics (ICDL), pp. 1–6 (2012)
4. Teulière, C., Forestier, S., Lonini, L., Zhang, C., Zhao, Y., Shi, B.E., Triesch, J.: Self-calibrating smooth pursuit through active efficient coding. Robot. Auton. Syst. **71**, 3–12 (2015)
5. Lonini, L., Zhao, Y., Chandrashekhariah, P., Shi, B.E., Triesch, J.: Autonomous learning of active multi-scale binocular vision. In: IEEE International Conference on Development and Learning and Epigenetic Robotics (ICDL), pp. 1–6 (2013)
6. Zhang, C., Triesch, J., Shi, B.E.: An active-efficient-coding model of optokinetic nystagmus. J. Vision **16**(14), 10–10 (2016)
7. Triesch, J., Eckmann, S., Shi, B.E.: A computational model for the joint development of accommodation and vergence control. J. Vision **17**(10), 162–162 (2017)
8. Vikram, T., Teulière, C., Zhang, C., Shi, B.E., Triesch, J.: Autonomous learning of smooth pursuit and vergence through active efficient coding. In: IEEE International Conference on Development and Learning and Epigenetic Robotics (ICDL) (2014)
9. Chandrapala, T.N., Shi, B.E., Triesch, J.: On the utility of sparse neural representations in adaptive behaving agents. In: International Joint Conference on Neural Networks (IJCNN) (2015)
10. Lonini, L., Forestier, S., Teulière, C., Zhao, Y., Shi, B.E., Triesch, J.: Robust active binocular vision through intrinsically motivated learning. Front. Neurorobotics **7**, 20–20 (2013)

11. Freeman, R., Pettigrew, J.: Alteration of visual cortex from environmental asymmetries. J. Nature **246**, 359–360 (1973)
12. Tanaka, S., Ribot, J., Imamura, K., Tani, T.: Orientation-restricted continuous visual exposure induces marked reorganization of orientation maps in early life. Neuroimage **30**, 462477 (2006)
13. Tanaka, S., Tani, T., Ribot, J., OHashi, K., Imamura, K.: A postnatal critical period for orientation plasticity in the cat visual cortex. PLoS ONE **4**, e5380 (2009)
14. Hirsch, H.V.B., Spinelli, D.N.: Visual experience modifies distribution of horizontally and vertically oriented receptive fields in cats. Science **168**, 869–871 (1970)
15. Wiesel, T.N., Hubel, D.H.: Single-cell responses in striate cortex of kittens deprived of vision in one eye. J. Neurophysiol. **26**, 10031017 (1963)
16. Hunt, J.J., Dayan, P., Goodhill, G.J.: Sparse coding can predict primary visual cortex receptive field changes induced by abnormal visual input. PLoS Comput. Biol. **9**(5), e1003005 (2013)
17. Klimmasch, L., Lelais, A., Lichtenstein, A., Shi, B.E., Triesch, J.: Learning of active binocular vision in a biomechanical model of the oculomotor system. bioRxiv 160721 (2017). https://doi.org/10.1101/160721
18. Priamikov, A., Fronius, M., Shi, B.E., Triesch, J.: OpenEyeSim: a biomechanical model for simulation of closed-loop visual perception. J. Vision **16**(15), 25–25 (2016)
19. Olmos, A., Kingdom, F.A.: A biologically inspired algorithm for the recovery of shading and reflectance images. Perception **33**(12), 1463–1473 (2004)
20. Mallat, S.G., Zhang, Z.: Matching pursuits with time-frequency dictionaries. IEEE Trans. Signal Process. **41**, 3397–3415 (1993)
21. Albert, M.V., Schnabel, A., Field, D.J.: Innate visual learning through spontaneous activity patterns. PLoS Comput. Biol. **4**(8), e1000137 (2008)
22. Sutton, R.S., Barto, A.G.: Reinforcement Learning: An Introduction. MIT Press, Cambridge (1998)
23. Dayan, P., Abbott, L.F.: Theoretical Neuroscience: Computational and Mathematical Modeling of Neural Systems. MIT Press, Cambridge (2005)
24. Van Hasselt, H., Wiering, M.A.: Reinforcement learning in continuous action spaces. In: IEEE International Symposium on Approximate Dynamic Programming and Reinforcement Learning, pp. 272–279 (2007)
25. Chandrapala, T.N., Shi, B.E., Triesch, J.: Active maintenance of binocular correspondence leads to orientation alignment of visual receptive fields. In: Joint IEEE International Conference on Development and Learning and Epigenetic Robotics (ICDL), pp. 98–103 (2015)
26. Appelle, S.: Perception and discrimination as a function of stimulus orientation: the "oblique effect" in man and animals. Psychol. Bull. **78**(4), 266–278 (1972)
27. Priamikov, A., Narayan, V., Shi, B.E., Triesch, J.: The role of contrast sensitivity in the development of binocular vision: A computational study. In: Joint IEEE International Conference on Development and Learning and Epigenetic Robotics (ICDL), pp. 33–38 (2015)
28. Leventhal, A.G., Hirsch, H.V.: Cortical effect of early selective exposure to diagonal lines. Science **190**(4217), S.902–S.904 (1975)
29. Stryker, M.P., Sherk, H., Leventhal, A.G., Hirsch, H.V.: Physiological consequences for the cat's visual cortex of effectively restricting early visual experience with oriented contours. J. Neurophysiology **41**(4), 896–909 (1978)

Learning Hierarchical Integration of Foveal and Peripheral Vision for Vergence Control by Active Efficient Coding

Zhetuo Zhao[1], Jochen Triesch[2], and Bertram E. Shi[3(✉)]

[1] Department of Brain and Cognitive Science, University of Rochester, Rochester, NY, USA
[2] Frankfurt Institute for Advanced Studies, Frankfurt am Main, Germany
[3] Department of Electronic and Computer Engineering,
Hong Kong University of Science and Technology, Kowloon, Hong Kong
eebert@ust.hk

Abstract. The active efficient coding (AEC) framework parsimoniously explains the joint development of visual processing and eye movements, e.g., the emergence of binocular disparity selective neurons and fusional vergence, the disjunctive eye movements that align left and right eye images. Vergence can be driven by information in both the fovea and periphery, which play complementary roles. The high resolution fovea can drive precise short range movements. The lower resolution periphery supports coarser long range movements. The fovea and periphery may also contain conflicting information, e.g. due to objects at different depths. While past AEC models did integrate peripheral and foveal information, they did not explicitly take into account these characteristics. We propose here a two-level hierarchical approach that does. The bottom level generates different vergence actions from foveal and peripheral regions. The top level selects one. We demonstrate that the hierarchical approach performs better than prior approaches in realistic environments, exhibiting better alignment and less oscillation.

Keywords: Vergence · Disparity · Foveal vision · Peripheral vision

1 Introduction

The perception-action cycle is at play during vergence eye movements. Depth information is encoded as the population responses of binocular disparity selective simple and complex cells in the primary visual cortex [1–3]. The population responses of these neurons drive the eye muscles via the ocular motor neurons so that the two eyes converge or diverge to align the left and right foveal images.

The fovea and the periphery play complementary roles in vergence control. Foveal disparity selective cells have small receptive field sizes [4] and high preferred spatial frequencies [5]. They provide precise disparity detection over small

© Springer Nature Switzerland AG 2018
P. Manoonpong et al. (Eds.): SAB 2018, LNAI 10994, pp. 78–89, 2018.
https://doi.org/10.1007/978-3-319-97628-0_7

ranges [6]. Peripheral disparity selective cells have larger receptive field sizes and lower preferred spatial frequencies. They provide robust, but less precise detection, over larger disparity ranges.

Fusional vergence can zero out initial retinal disparities of up to four degrees [7,8], while the fovea covers a region only one degree in diameter. This suggests that fusional vergence control requires cooperation between the fovea and periphery. Information in the periphery is useful during the initial stages of vergence, when disparities may be large. On the other hand, when the foveal images are aligned, objects located at other depths may result in nonzero disparities in the periphery, suggesting that some information in the periphery should be ignored. Tanimoto showed that disturbances presented in the periphery influenced the vergence latency, but not the steady state amplitude of vergence angle [30], suggesting that peripheral vision is involved in the early stages of vergence, but not at steady state.

This paper proposes a mechanism that enables an agent to learn how to both process and integrate visual information from the fovea and periphery in order to achieve robust vergence control. Much past work in neuromorphic vergence control either did not address the problem of learning and relied upon hand crafted perceptual and control strategies [9–12], or studied learning of only one aspect, i.e. control [13–15].

The mechanism is based upon the active efficient coding (AEC) framework, which extends the efficient coding hypothesis [16,17] to include action. In contrast to past work on joint learning of vergence [18,19], the AEC does not require a fixed pre-defined reward signal. Prior work applying AEC to this problem [20–22]. did not address how information from the fovea and periphery could be integrated differently at different stages of vergence.

Here, we present a hierarchical mechanism inspired by [23–25] that addresses this problem. Different regions of the visual field first generate different actions. One of these commands is then selected based upon the visual information. The usage of information in the fovea and periphery changes dynamically in a stimulus dependent manner. Our experimental results demonstrate that this leads to more robust vergence in complex environments.

2   Models

Figure 1 shows a block diagram of our system for the joint development of disparity perception and fusional vergence behavior. It consists of three components: (1) the pre-processing component, which extracts stereo image patches from sub-windows of the original stereo-images; (2) the perceptual component, which encodes the stereo patches as the responses of a set of disparity selective units using a GASSOM encoder [26]; (3) the behavior component, which maps the output of perceptual component to a vergence action via a neural network. The perceptual and behavioral components are learned simultaneously as the agent behaves in the environment, through unsupervised and reinforcement learning respectively.

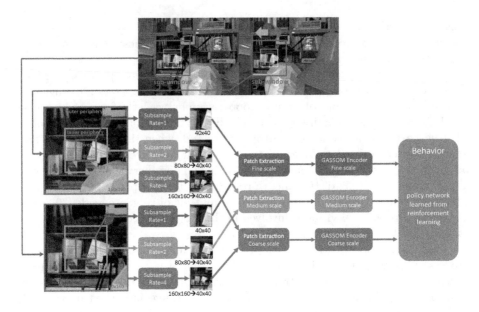

Fig. 1. Architecture of the vergence joint development system.

2.1 Patch Extraction

We extract three subwindows from each image: a coarse scale subwindow, a medium scale subwindow and a fine scale subwindow, Following a pyramidal architecture, the coarse (medium) scale subwindow is four (two) times the size of the fine scale subwindow, but is downsampled by a factor of four (two), so that all subwindows have the same size (40 by 40 pixels). These subwindows are further subdivided into a 7 by 7 array of 10 by 10 pixel patches with a stride of 5 pixels. As shown in Fig. 2, patches from the different scales correspond to different sized regions in the original image.

We index scale by $s \in \{f, m, c\}$ (fine, medium, coarse) and patch location by $i, j \in \{-3, \ldots, 0, \ldots, 3\}$. We create stereo image patches $\mathbf{x}_{s,i,j} \in \mathbb{R}^{200 \times 1}$ by combining patches from the two eyes.

$$\mathbf{x}_{s,i,j} = \begin{bmatrix} \mathbf{x}_{s,i,j}^{\mathrm{L}} \\ \mathbf{x}_{s,i,j}^{\mathrm{R}} \end{bmatrix} \tag{1}$$

where $\mathbf{x}_{s,i,j}^{\mathrm{L}}, \mathbf{x}_{s,i,j}^{\mathrm{R}} \in \mathbb{R}^{100 \times 1}$ represent left and right monocular patches.

For intensity and contrast invariance, we apply mean subtraction so that $\mathbf{x}_{s,i,j}^{\mathrm{L}}$ and $\mathbf{x}_{s,i,j}^{\mathrm{R}}$ individually have zero mean, followed by normalization so that $\mathbf{x}_{s,i,j}$ has unit variance.

2.2 Perception

We use the Generative Adaptive Subspace Self Organizing Map (GASSOM) [26] to generate a perceptual representation of the stereo image patches.

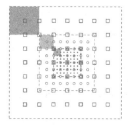

Fig. 2. The pyramid structure of the subwindows and patches. The dotted lines show boundaries of the subwindows for the fine (blue), medium (green) and coarse (red) scale subwindows. The solid squares show the size of the region in the original image covered by a single patch. The empty squares show the center locations of the patches. (Color figure online)

The GASSOM algorithm assumes that the high ($N = 200$) dimensional visual input is generated by sampling from a lower ($B = 2$) dimensional subspace chosen from a dictionary of $K = 324$ subspaces. The subspaces are learned through exposure to unlabeled image patches via an unsupervised learning algorithm. The learned subspaces represent patches with different oriented textures and stereo disparities.

Patches in the same scale s share the same dictionary. Each subspace in the dictionary is spanned by a pair of orthogonal basis vectors contained in the columns of the matrix $\mathbf{B}_{s,k} = \begin{bmatrix} \mathbf{b}_{s,k,1} & \mathbf{b}_{s,k,2} \end{bmatrix} \in \mathbb{R}^{200 \times 2}$ where $k \in \{0, 1, \ldots, 323\}$ indexes the dictionary elements.

The squared length of a binocular input patch's projection onto the subspace spanned by $\mathbf{B}_{s,k}$ is a measure of the extent to which the subspace can account for the input patch. It is also similar to the disparity energy [1], which is often used to model the output of disparity selective binocular complex cells. The disparity is the sum of the squared outputs of two linear binocular neurons. Because the two columns are orthogonal, the squared length of the projection can be calculated by $||\mathbf{B}_{s,k}^T \mathbf{x}_{s,i,j}|| = (\mathbf{b}_{s,k,1}^T \mathbf{x}_{s,i,j})^2 + (\mathbf{b}_{s,k,2}^T \mathbf{x}_{s,i,j})^2$. Thus, each basis vector is analogous to the receptive field of a linear binocular neuron. Each basis vector can be split into two parts, e.g. $\mathbf{b}_{s,k,1}^L, \mathbf{b}_{s,k,1}^R \in \mathbb{R}^{100}$, one corresponding to the left eye and one corresponding to the right eye. Each part can be rearranged into a 10×10 matrix.

The basis vectors are learned as the agent behaves in the environment by an unsupervised learning procedure [26]. The basis vectors of each subspace k are updated in the direction that minimizes the reconstruction error of the subspace, $||\mathbf{x}_{s,i,j} - \mathbf{B}_{s,k} \mathbf{B}_{s,k}^T \mathbf{x}_{s,i,j}||^2$. The reconstruction error is a the squared distance between the input vector and its projection onto the subspace. The size of the update depends upon how likely the subspace accounts for the input vector.

After learning, the basis vectors develop so that the 10×10 matrices have Gabor-like profiles with similar spatial frequencies and orientations [20,21]. For each eye, two basis functions, e.g. $\mathbf{b}_{s,k,1}^L$ and $\mathbf{b}_{s,k,2}^L$ are in approximate phase quadrature. For each basis vector, the left and right eye components, e.g. $\mathbf{b}^L_{s,k,1}$ and $\mathbf{b}^R_{s,k,1}$ have a phase shift that determines the preferred disparity of the subspace. Thus, the learned basis vectors have properties similar to the linear simple cell binocular receptive fields in the disparity energy model.

For each scale s and each patch i, j, we define an output feature vector $\mathbf{c}_{s,i,j} \in \mathbb{R}^{324}$ as the set of squared projections onto all subspaces:

$$
\mathbf{c}_{s,i,j} = \begin{bmatrix} ||\mathbf{B}^T_{s,0}\mathbf{x}_{s,i,j}||^2 \\ \vdots \\ ||\mathbf{B}^T_{s,324}\mathbf{x}_{s,i,j}||^2 \end{bmatrix} \tag{2}
$$

This feature vector models the population output of a set of disparity, spatial frequency and orientation selective complex cells in the primary visual cortex all serving the same location.

2.3   Behavior

Behavior is defined by a policy that maps the output of the perceptual representation to a vergence action, We simulate vergence by changing the horizontal shift between the center locations of the subwindows extracted from the left and right eye images. The vergence actions used here update this shift by an amount $A \in \{-16, -8, -4, -2, -1, 0, 1, 2, 4, 8, 16\}$ in pixels.

Below, we describe two neural network based architectures for mapping the perceptual representations to vergence actions: a parallel model, which is the same as used in prior work [20, 21] and a new hierarchical model.

Parallel Model. The parallel model (shown in Fig. 3(a)) combines the population outputs $\mathbf{c}_{s,i,j} \in \mathbb{R}^{324}$ in (2) into one long feature vector $\mathbf{F}_P \in \mathbb{R}^{972}$, by pooling across space and concatenating across scale

$$
\mathbf{F}_P = \begin{bmatrix} \frac{1}{7^2} \sum_{i,j=-3}^{3} \mathbf{c}_{c,i,j} \\ \frac{1}{7^2} \sum_{i,j=-3}^{3} \mathbf{c}_{m,i,j} \\ \frac{1}{7^2} \sum_{i,j=-3}^{3} \mathbf{c}_{f,i,j} \end{bmatrix} \tag{3}
$$

A single layer neural network with a softmax output nonlinearity maps this feature vector to a probability distribution over the 11 actions, $\mathbf{p} = [p_i]_{i=0}^{10} \in \mathbb{R}^{11}$:

$$
\mathbf{p} = \text{softmax}(\mathbf{a}) \tag{4}
$$

where $\mathbf{a} = [a_i]_{i=0}^{10} \in \mathbb{R}^{11}$ is a set of motor neuron outputs computed by

$$
\mathbf{a} = \mathbf{W}_P \cdot \mathbf{F}_P \tag{5}
$$

where $\mathbf{W}_P \in \mathbb{R}^{11 \times 972}$. The softmax operator is given by $p_i = e^{a_i/T}(\sum_{j=0}^{N} e^{a_j/T})^{-1}$, where T is a temperature parameter.

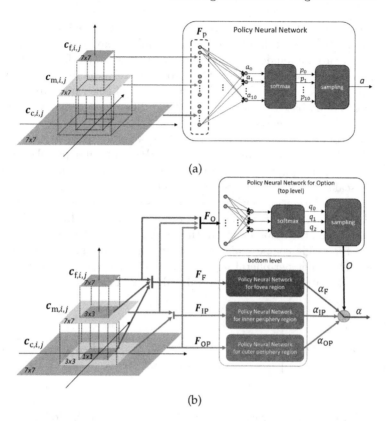

(a)

(b)

Fig. 3. (a) Architecture of the parallel model. (b) Architecture of the hierarchical model.

The weights $\mathbf{W_P}$ develop following the Natural Actor-Critic Reinforcement Learning (NACREL) algorithm [27] to maximize the discounted sum of instantaneous rewards. The instantaneous reward is the reconstruction error of best matching subspace $e_{s,i,j}$ averaged across all scales and all rewards

$$r_P = -\frac{1}{3 \times 7^2} \sum_s \sum_{i,j} e_{s,i,j} \tag{6}$$

where

$$e_{s,i,j} = ||\mathbf{x}_{s,i,j} - \mathbf{B}_{s,\hat{k}} \mathbf{B}_{s,\hat{k}}^T \mathbf{x}_{s,i,j}||^2 \tag{7}$$

and $\hat{k} = \mathrm{argmax}_k ||\mathbf{B}_{s,k}^T \mathbf{x}_{s,i,j}||^2$. The summation ranges over $s \in \{f,m,c\}$ and $i,j \in \{-3,\ldots,3\}$.

Note that both the perceptual component and the behavioral component develop so as to minimize the reconstruction error. This shared goal ensures the stability and robustness of a learning process where both perception and behavior are evolving simultaneously.

Hierarchical Model. While the parallel model works well with large planar objects, it does not perform well in more realistic environments, where the objects being fixated upon do not cover the entire extent of the fine, medium and coarse scale subwindows. To address this problem, we propose here a two level hierarchical model shown in Fig. 3B. The bottom level generates three separate vergence commands based on information from different subwindows. The top level selects one of these commands.

At the bottom level, the system defines three separate input feature vectors, which we label as foveal (F), inner peripheral (IP), and outer peripheral (OP). These inputs gather inputs from the image regions covered by the fine, medium and coarse scale subwindows respectively, but may contain information from multiple scales. Information within each scale is combined by spatial pooling. The foveal input, $\mathbf{F}_{\mathrm{F}} \in \mathbb{R}^{972}$, depends not only the responses from all fine scale patches, but also on the responses from the three by three array of medium scale patches and the single coarse scale patch that fall entirely inside the fine scale subwindow. The inner peripheral input, $\mathbf{F}_{\mathrm{IP}} \in \mathbb{R}^{648}$, depends on both medium and coarse scale patches. The outer peripheral input, $\mathbf{F}_{\mathrm{OP}} \in \mathbb{R}^{324}$, depends only coarse scale input.

$$\mathbf{F}_{\mathrm{F}} = \begin{bmatrix} \mathbf{c}_{\mathrm{c},0,0} \\ \frac{1}{3^2} \sum_{i,j=-1}^{1} \mathbf{c}_{\mathrm{m},i,j} \\ \frac{1}{7^2} \sum_{i,j=-3}^{3} \mathbf{c}_{\mathrm{f},i,j} \end{bmatrix} \quad \mathbf{F}_{\mathrm{IP}} = \begin{bmatrix} \frac{1}{3^2} \sum_{i,j=-1}^{1} \mathbf{c}_{\mathrm{c},i,j} \\ \frac{1}{7^2} \sum_{i,j=-3}^{3} \mathbf{c}_{\mathrm{m},i,j} \end{bmatrix} \quad \mathbf{F}_{\mathrm{OP}} = \frac{1}{7^2} \sum_{i,j=-3}^{3} \mathbf{c}_{\mathrm{c},i,j}$$

$$(8)$$

Each feature vector generates a different probability distribution over actions following a similar strategy as used in the parallel model, i.e. equations (4) and (5) with appropriate changes in the dimensionality of the input feature vector and weight matrix.

Weights were learned using the NACREL algorithm. The instantaneous reward for each networks was the average reconstruction errors of the patches included in its feature vector.

$$r_{\mathrm{F}} = -\frac{1}{3} \left(e_{\mathrm{c},0,0} + \frac{1}{3^2} \sum_{i,j=-1}^{1} e_{\mathrm{m},i,j} + \frac{1}{7^2} \sum_{i,j=-3}^{3} e_{\mathrm{f},i,j} \right) \qquad (9)$$

$$r_{\mathrm{IP}} = -\frac{1}{2} \left(\frac{1}{3^2} \sum_{i,j=-1}^{1} e_{\mathrm{c},i,j} + \frac{1}{7^2} \sum_{i,j=-3}^{3} e_{\mathrm{m},i,j} \right) \quad r_{\mathrm{OP}} = -\frac{1}{7^2} \sum_{i,j=-3}^{3} e_{\mathrm{c},i,j} \quad (10)$$

The top level, selects from among the three options based on the pooled information from all scales, $\mathbf{F}_{\mathrm{O}} = \mathbf{F}_{\mathrm{P}}$, using a neural network with the same structure as the parallel network, except the dimensionality of the output vector was three. Each output corresponds to the selection of the vergence command from one of the bottom layer networks. The weights optimized the same reward as the parallel model, $r_{\mathrm{O}} = r_{\mathrm{P}}$.

3   Experimental Procedure

A simulated agent was presented with a sequence of virtual scenes. Within each scene, the agent executed saccades to 20 different fixation locations, chosen randomly by sampling from the saliency distribution computed on the left eye image using Attention based on Information Maximization (AIM) [28]. At each fixation location, the agent executed 10 vergence commands. After 20 fixations, a new scene was presented to the agent.

During training, vergence actions α were generated by sampling from the multinomial probability distribution specified by $\mathbf{p}$. During testing, actions were generated using a greedy policy, which chose the most likely action.

We mimic the effect of eye movements in a scene by changing the center locations of the subwindows taken from a single stereo image pair from the Tsukuba Stereo Dataset [29]. This dataset includes 1800 stereo pairs generated by rendering the video obtained by a parallel stereo camera pair moving through a simulated laboratory environment and corresponding ground truth disparity maps. Since adjacent frames in the video are very similar, we took one out of every five frames in the sequence, resulting in 360 stereo images. Twenty were selected randomly as the testing set. The remainder were used as the training set.

Fixation/vergence actions were executed by setting the center location of the left eye subwindows equal to the chosen fixation point. The center position of the right eye subwindows was offset horizontally by the vergence, which was updated according the vergence actions chosen.

Due to the parallel camera geometry of the Tsukuba dataset and the way we simulated vergence movements, the system only encounters horizontal disparities. In reality, a convergent camera geometry will introduce vertical disparities due to shifts in the epipolar lines. However, we do not expect this additional complexity to dramatically alter the results presented here. The AEC framework leads to similar results for vergence control using eye movements simulated by window movement [20] and for camera movements in simulated and real environments [22].

4   Experimental Results

4.1   Trained Policies

Figure 4 visualizes the policies learned during one trial of the hierarchical and parallel models as images. Each column shows the value of $\mathbf{p}$ in (4) at a particular disparity d averaged over 100 binocular inputs generated from randomly selected left eye images in the test set. Left eye subwindows were taken around a randomly generated center location. Right eye subwindows were taken from the same image around a center location offset by d. This ensured that the disparity is uniform across the subwindows.

The learned policies have properties that reflect and appropriately exploit the properties of the different inputs.

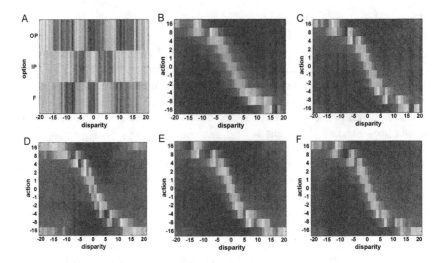

Fig. 4. Learned policies from one trial. Each column of an image shows the probability of an action given the input disparity using the jet colormap, which varies from blue (0) to red (1). (A-E) Hierarchical model: (A) top level, (B) outer peripheral, (C) inner peripheral, (D) foveal, (E) combined. (F) Parallel model. (Color figure online)

The top level policy of the hierarchical model (Fig. 4A) tends to choose the action generated by the outer peripheral policy when the disparity is large, by the foveal policy when the disparity is small, and the inner peripheral policy for intermediate disparities, leading to a "V" shaped image.

For the vergence action generating policies (Fig. 4B–F), we expect to see an upside-down sigmoid due to the exponential action spacing. Positive input disparities generate negative vergence actions to zero-out the disparity. For the hierarchical model, the outer peripheral policy (B) reliably generates large vergence actions for large input disparities, but is less reliable at small input disparities, due to the coarser resolution. In contrast, the foveal policy (D) more reliably generates actions that precisely cancel the input disparity when it is small, but is less reliable at larger disparities, due to the limited spatial extent of the patches (10 by 10 pixels). When the input disparity is near −20 or +20 pixels, we observe a bimodal distribution with centers at the −10 and +10 vergence actions. The left and right eye patches image non-overlapping regions in the environment. The policy reliably detects that the input disparity is large, but is unsure about the direction to change the vergence. The inner peripheral policy (C) exhibits intermediate characteristics.

We estimated the policy combining the top and bottom levels by averaging each column of the three vergence action policies weighted by the probability that each policy was selected at the top level. The combined policy (E) is very similar to the policy from the parallel method (F), as we might expect for these inputs where the disparity is uniform over all subwindows.

4.2   Vergence in Complex Environments

The difference between the parallel and hierarchical model is clearly evident in more complex environments where the disparities encountered in the fovea and periphery may differ, e.g. when the agent is fixating on a small object that does not cover the entire periphery and the background is at a different depth.

Figure 5A shows the vergence angle trajectories generated by the parallel and hierarchical models for a sequence of seven fixations on one of the stereo images in the test set. The parallel model failed to converge to the correct vergence angle at the 2nd, 5th and 7th fixations, and exhibited oscillatory behavior due to conflicting disparities in the fovea and periphery. The hierarchical model performed much better, converging without oscillation on all fixations.

Figure 5B shows that at the beginning of the fixations, when the retinal disparities are large, the hierarchical model tends to choose actions generated by the peripheral regions. At the end of the fixations, when the retinal disparity at the fixation point is small, it chooses actions generated by the fovea. A clear progression from outer periphery to fovea can be seen in the 5th and 6th fixations.

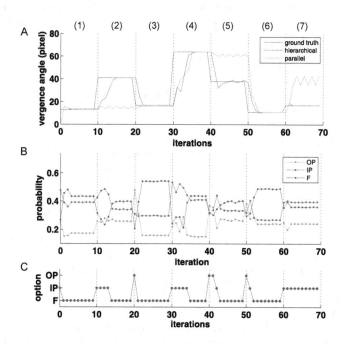

Fig. 5. Model outputs over 7 fixations. (A) The vergence angle trajectories generated by the hierarchical (red) and parallel (green) models in comparison to the ground truth (blue) vergence angles required to achieve zero retinal disparity at the subwindow center. (B) The probabilities for selecting the actions generated from the outer (green) and inner (red) peripheral and foveal (blue) networks, as computed by the top level network. (C) The option chosen by the top level network. Dotted gray lines delineate fixations, which are numbered for reference at the top. (Color figure online)

5   Conclusion

We have proposed a computational architecture for the joint learning of disparity perception and vergence control, which incorporates a hierarchical model for integrating information from the fovea and periphery. This architecture enables an agent to learn how to resolve conflicting disparities in different image regions through its interaction with the environment, and without an explicit teaching signal. The policy that emerges exhibits behavior reminiscent of coarse to fine behavior.

Acknowledgements. This work was supported by the Hong Kong Research Grants Council under Grant 16244416, the German Federal Ministry of Education and Research under Grants 01GQ1414 and 01EW1603A, the European Union's Horizon 2020 Grant 713010, and the Quandt Foundation.

References

1. Ohzawa, I., DeAngelis, G., Freeman, R.: Stereoscopic depth discrimination in the visual cortex: neurons ideally suited as disparity detectors. Science **249**(4972), 1037–1041 (1990)
2. Hubel, D.H., Wiesel, T.N.: Receptive fields of single neurones in the cats striate cortex. J. Physiol. **148**(3), 574–591 (1959)
3. Hubel, D.H., Wiesel, T.N.: Receptive fields, binocular interaction and functional architecture in the cats visual cortex. J. Physiol. **160**(1), 106–154 (1962)
4. Freeman, J., Simoncelli, E.P.: Metamers of the ventral stream. Nat. Neurosci. **14**(9), 1195–1201 (2011)
5. Henriksson, L., Nurminen, L., Hyvarinen, A., Vanni, S., Hyvarinen, A.: Spatial frequency tuning in human retinotopic visual areas. J. Vision **8**(10), 1–13 (2008)
6. Rawlings, S.C., Shipley, T.: Stereoscopic acuity and horizontal angular distance from fixation. J. Opt. Soc. Am. **59**(8), 991–993 (1969)
7. Antona, B., Barrio, A., Barra, F., Gonzalez, E., Sanchez, I.: Repeatability and agreement in the measurement of horizontal fusional vergences. Ophthalmic Physiol. Opt. **28**(5), 475–491 (2008)
8. Stevenson, S.B., Reed, P.E., Yang, J.: The effect of target size and eccentricity on reflex disparity vergence. Vision. Res. **39**(4), 823–832 (1999)
9. Siebert, J., Wilson, D.: Foveated vergence and stereo. In: Proceedings of the 3rd International Conference on Visual Search (1992)
10. Westelius, C., Knutsson, H., Wiklund, J., et al.: 11. Phase-Based Disparity Estimation. Vision Process Basic Res. Comput. Vision Syst. **10** (1995)
11. Gibaldi, A., Chessa, M., Canessa, A., et al.: A cortical model for binocular vergence control without explicit calculation of disparity. Neurocomputing **73**, 1065–1073 (2010)
12. Zhang, X., Tay, L.P.: Binocular vergence control using disparity energy neurons. J. Exp. Theor. Artif. Intell. **23**, 201–222 (2011)
13. Piater, J. H., Grupen, R. A., Ramamritham, K.: Learning real-time stereo vergence control. In: Proceedings of the 1999 IEEE International Symposium on Intelligent Control/Intelligent Systems and Semiotics, pp. 272–277. IEEE (1999)

14. Gibaldi, A., Canessa, A., Chessa, M., et al.: How a population-based representation of binocular visual signal can intrinsically mediate autonomous learning of vergence control. Procedia Comput. Sci. **13**, 212–221 (2012)

15. Gibaldi, A., Canessa, A., Solari, F., et al.: Autonomous learning of disparity-vergence behavior through distributed coding and population reward: basic mechanisms and real-world conditioning on a robot stereo head. Robot. Autonom. Syst. **71**, 23–34 (2015)

16. Barlow, H.: Possible principles underlying the transformations of sensory messages. Sens. Commun. **6**(2), 57–58 (1961)

17. Blattler, F., Hahnloser, R.H.R.: An efficient coding hypothesis links sparsity and selectivity of neural responses. PLoS ONE **6**(10), e25506 (2011)

18. Franz, A., Triesch, J.: Emergence of disparity tuning during the development of vergence eye movements. In: IEEE 6th International Conference on Development and Learning, pp. 31–36. IEEE (2007)

19. Sun, W., Shi, B. E.: Joint development of disparity tuning and vergence control. In: IEEE International Conference on Development and Learning (2011)

20. Zhao, Y., Rothkopf, C.A., Triesch, J., Shi, B.E.: A unified model of the joint development of disparity selectivity and vergence control. In: 2012 IEEE International Conference on Development and Learning and Epigenetic Robotics. IEEE (2012)

21. Lonini, L., Zhao, Y., Chandrashekhariah, P., Shi, B.E., Triesch, J.: Autonomous learning of active multi-scale binocular vision. In: 2013 IEEE 3rd Joint International Conference on Development and Learning and Epigenetic Robotics. IEEE (2013)

22. Lonini, L., Forestier, S., Teulire, C., et al.: Robust active binocular vision through intrinsically motivated learning. Front. Neurorobotics **7** (2013)

23. Sutton, R.S., Precup, D., Singh, S.: Between MDPs and semi-MDPs: a framework for temporal abstraction in reinforcement learning. Artif. Intell. **112**(1), 181–211 (1999)

24. Barto, A.G., Mahadevan, S.: Recent advances in hierarchical reinforcement learning. Disc. Event Dyn. Syst. **13**(12), 341–379 (2003)

25. Dietterich, T.G.: Hierarchical reinforcement learning with the MAXQ value function decomposition. J. Artif. Intell. Res. **13**, 227–303 (2000)

26. Chandrapala, T.N., Shi, B.E.: Learning Slowness in a Sparse Model of Invariant Feature Detection. Neural Comput. (2015)

27. Bhatnagar, S., Sutton, R.S., Ghavamzadeh, M., Lee, M.: Incremental natural actor-critic algorithms. Automatica **45**(11), 2471–2482 (2009)

28. Bruce, N.D.B., Tsotsos, J.K.: Saliency, attention, and visual search: an information theoretic approach. J. Vision **3**(9), 1–24 (2009)

29. Martull, S., Peris, M., Fukui, K.: Realistic CG stereo image dataset with ground truth disparity maps. In: International Conference on Pattern Recognition, pp. 40–43 (2012)

30. Tanimoto, N., Takagi, M., Bando, T., Abe, H., Hasegawa, S., Usui, T., Zee, D.S.: Central and peripheral visual interactions in disparityinduced vergence eye movements: I. Spatial interaction. Invest. Ophthalmol. Vis. Sci. **45**(4), 1132–1138 (2004)

Decentralized Control Scheme for Coupling Between Undulatory and Peristaltic Locomotion

Takeshi Kano$^{(\boxtimes)}$, Naoki Matsui, and Akio Ishiguro

Research Institute of Electrical Communication, Tohoku University,
2-1-1 Katahira, Aoba-ku, Sendai 980-8577, Japan
{tkano,n-matsui,ishiguro}@riec.tohoku.ac.jp
http://www.cmplx.riec.tohoku.ac.jp/

Abstract. Animals use various locomotion patterns, such as undulatory, peristaltic, and legged locomotion, reasonably and appropriately to adapt to a wide range of environments. The goal of this study is to design a decentralized control scheme that can generate multiple locomotion patterns in response to environments. We draw inspiration from our previous works on snake and earthworm locomotion and propose a decentralized control scheme capable of generating undulatory and peristaltic locomotion. We demonstrate via simulations that undulatory and peristaltic locomotion are used appropriately to adapt to various environments in real time.

Keywords: Decentralized control · Undulatory locomotion
Peristaltic locomotion

1 Introduction

Animals exhibit astoundingly adaptive locomotion under unpredictable and unstructured real-world environments. Interestingly, many animals possess various locomotion patterns, such as undulatory, peristaltic, and legged locomotion, and use them reasonably and appropriately to adapt to a wide range of environments (Fig. 1). For example, salamanders swim in water by undulating their body trunk while they walk on the land by using their legs [1]. This ability has been honed by evolutionary selection pressure, and it is likely that an ingenious underlying mechanism exists that leads to highly adaptive behavior. Robots would be able to perform effectively in various environments if this remarkable mechanism could be applied to them.

However, current bioinspired robots [2–5] are still far less adaptive than animals. This is because most of the previous studies on bioinspired robotics focused on individual locomotion patterns. Although a few studies focused on the transition between different locomotion patterns [1,6], adaptability was still limited, because sensory feedback mechanisms that could lead to generation of locomotion patterns reasonable for the environment were not implemented.

© Springer Nature Switzerland AG 2018
P. Manoonpong et al. (Eds.): SAB 2018, LNAI 10994, pp. 90–101, 2018.
https://doi.org/10.1007/978-3-319-97628-0_8

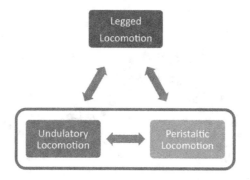

Fig. 1. Schematic of locomotion patterns. This study focuses on the coupling between the undulatory and peristaltic locomotion.

In this study, we aim to design a decentralized control scheme that can generate multiple locomotion patterns in response to different environments. As a first step, we focus on undulatory and peristaltic locomotion and consider how to couple them reasonably. We have previously proposed a decentralized control scheme for peristaltic locomotion of earthworms [7] and undulatory locomotion of snakes [8]. On this basis, we propose a decentralized control scheme for a robot in which several segments capable of bending, contracting, and elongating are concatenated one dimensionally. We demonstrate via simulations that the robot can exhibit both undulatory and peristaltic locomotion to adapt to various environments in real time.

2   Model

2.1   Mechanical System

The mechanical system is described by a simple two-dimensional mass-spring-damper model, as shown in Fig. 2. Masses that represent the body trunk are connected one-dimensionally via the parallel combination of a damper and a real-time tunable spring (RTS), which is a spring whose natural length can be actively changed [9]. Masses that represent the body walls on both sides are connected to adjacent body trunk masses via springs and dampers. Adjacent body wall masses are connected by the parallel combination of a damper and an RTS. Torsional springs are embedded in the body trunk masses to prevent excessive bending of the body. The natural lengths of the RTSs of the body trunk, right body wall, and left body wall are denoted by $\bar{l}_{c,i}$, $\bar{l}_{r,i-\frac{1}{2}}$, and $\bar{l}_{l,i-\frac{1}{2}}$, respectively, where their subscripts are defined as shown in Fig. 2. A segment contracts (expands) when the natural lengths of all RTSs at the corresponding segment decrease (increase) simultaneously, whereas it bends when there is a difference between the natural lengths of the RTSs of the right and left body walls.

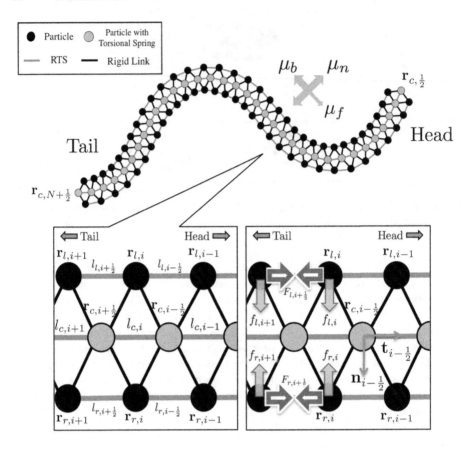

Fig. 2. Schematic of the model.

Normal forces acting on the body wall, the real lengths of the RTSs, and the forces generated at the RTSs can be obtained as sensory information.

For simplicity, we assume that the friction coefficient of the body wall masses is zero and that the ground frictional forces act only on the body trunk masses. The friction coefficient of the body trunk masses is designed according to the following two rules:

Rule 1. The friction coefficient increases as the body segment contracts, as with earthworms.

Rule 2. The friction coefficient in the forward direction is smaller than those in the lateral and backward directions, as with snakes.

Based on the above rules, the frictional force is modeled by Coulomb friction. However, it is difficult to reproduce static friction in simulations. Hence, the friction force vector acting on the $i + \frac{1}{2}$th body trunk mass $\mathbf{f}_{fric,i+\frac{1}{2}}$ is simply described by the following equation:

$$\mathbf{f}_{fric,i+\frac{1}{2}} = -\underbrace{\frac{mg}{(l_{c,i} + l_{c,i+1})^n}}_{\text{Rule1.}} \Big\{ \underbrace{\mu_f \tanh(\max[\lambda \dot{\mathbf{r}}_{c,i+\frac{1}{2}} \cdot \mathbf{t}_{c,i+\frac{1}{2}}, 0]) \mathbf{t}_{c,i+\frac{1}{2}}}_{\text{Rule2.}}$$

$$\underbrace{+ \mu_b \tanh(\min[\lambda \dot{\mathbf{r}}_{c,i+\frac{1}{2}} \cdot \mathbf{t}_{c,i+\frac{1}{2}}, 0]) \mathbf{t}_{c,i+\frac{1}{2}}}_{\text{Rule2.}}$$

$$\underbrace{+ \mu_n \tanh(\lambda \dot{\mathbf{r}}_{c,i+\frac{1}{2}} \cdot \mathbf{n}_{c,i+\frac{1}{2}}) \mathbf{n}_{c,i+\frac{1}{2}} \Big\}}_{\text{Rule2.}}, \tag{1}$$

where m is the mass, g is the gravitational acceleration, μ_f, μ_b, μ_n, λ, and n are positive constants, with μ_f being smaller than μ_b and μ_n, $\mathbf{t}_{c,i+\frac{1}{2}}$ and $\mathbf{n}_{c,i+\frac{1}{2}}$ denote unit vectors tangential and normal to the body axis, respectively, $l_{c,i}$ denotes the actual length of the RTS at the center, and $\mathbf{r}_{c,i+\frac{1}{2}}$ denotes the position of the $i + \frac{1}{2}$th body trunk mass (Fig. 2). The terms corresponding to Rules 1 and 2 are shown below the equation.

2.2 Control System

The control scheme is designed based on our previous works on snake and earthworm locomotion [7,8]. The natural lengths of the RTSs for several segments from the head are controlled in a centralized manner, whereas those for the other segments are controlled in a decentralized manner so that the body reasonably bends, contracts, and expands in response to the environment. Although we do not describe explicitly in the equations below, the maximum and minimum natural lengths are defined for all RTSs to avoid excessive body deformation. Specifically, the natural length is reset to $\bar{l}_{max,p}$ and $\bar{l}_{min,p}$ if it is above $\bar{l}_{max,p}$ and below $\bar{l}_{min,p}$, respectively ($p = r, l$, and c, which represent right, left, and center, respectively).

Centralized Control of Head Segments. The RTSs on the first N_β segments are controlled as follows:

$$\bar{l}_{l,i+\frac{1}{2}}(t+1) = l_0, \tag{2}$$

$$\bar{l}_{r,i+\frac{1}{2}}(t+1) = l_0, \tag{3}$$

$$\bar{l}_{c,i}(t+1) = l_0 \qquad\qquad (0 < i \leq N_\alpha), \tag{4}$$

$$\bar{l}_{l,i+\frac{1}{2}}(t+1) = l_0 + h_u(t) + h_p(t), \tag{5}$$

$$\bar{l}_{r,i+\frac{1}{2}}(t+1) = l_0 - h_u(t) + h_p(t), \tag{6}$$

$$\bar{l}_{c,i}(t+1) = l_0 + h_p(t) \qquad\qquad (N_\alpha < i \leq N_\beta), \tag{7}$$

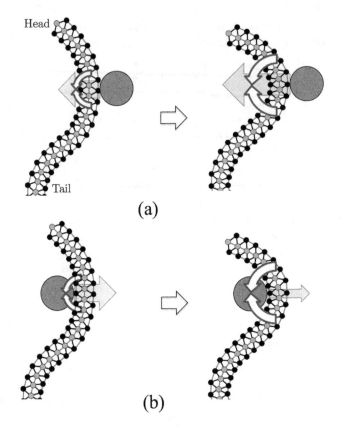

Head

Tail

(a)

(b)

Fig. 3. Mechanism of *Tegotae*-based control: (a) good *Tegotae* and (b) bad *Tegotae*.

where t denotes the time step, l_0 denotes the nominal length of the RTSs, and $h_u(t)$ and $h_p(t)$ denote the motor command for undulatory and peristaltic motion, respectively, from a higher center. The motor command is not applied to the first N_α segments, so that the head part does not get stuck in narrow aisles.

Decentralized Control for Bending. The natural length of the body wall masses for ($N_\beta < i \leq N$) is controlled according to the following equation:

$$\bar{l}_{l,i+\frac{1}{2}}(t+1) = l_{l,i-\frac{1}{2}}(t) + \sigma \tanh\left(-\kappa_u \sum_{k=i-n_b}^{i+n_f} T_{k+\frac{1}{2}}\right), \qquad (8)$$

$$\bar{l}_{r,i+\frac{1}{2}}(t+1) = l_{r,i-\frac{1}{2}}(t) + \sigma \tanh\left(\kappa_u \sum_{k=i-n_b}^{i+n_f} T_{k+\frac{1}{2}}\right)$$

$$(N_\beta < i < N). \qquad (9)$$

where

$$T_{i+\frac{1}{2}} = (F_{l,i+\frac{1}{2}}(t) - F_{r,i+\frac{1}{2}}(t))(\tilde{f}_{l,i+\frac{1}{2}}(t) + \tilde{f}_{r,i+\frac{1}{2}}(t)).$$ (10)

$$\tilde{f}_{l,i+\frac{1}{2}}(t) = \frac{f_{l,i} + f_{l,i+1}}{2},$$ (11)

$$\tilde{f}_{r,i+\frac{1}{2}}(t) = \frac{f_{r,i} + f_{r,i+1}}{2}.$$ (12)

with κ_u, σ, n_b, and n_f being positive constants. Here, $l_{l,i+\frac{1}{2}}$ and $l_{r,i+\frac{1}{2}}$ denote the actual length of the RTS on the left- and right-hand sides, respectively; $F_{l,i+\frac{1}{2}}$ and $F_{r,i+\frac{1}{2}}$ denote the force generated by the RTS on the left- and right-hand sides, respectively; and $f_{l,i}$ and $f_{r,i}$ denote the normal force acting on the ith body wall mass on the left- and right-hand sides, respectively.

The first terms on the right-hand side of Eqs. (8) and (9) correspond to curvature derivative control proposed previously as the control scheme for snakelike robots [10]. Owing to this term, bending torques are generated at the point where the curvature derivative of the body curve is large, and as a consequence, the body tends to follow a path established by the head.

The second terms on the right-hand side of Eqs. (8) and (9) denote local sensory feedback based on *Tegotae*. *Tegotae* is a subjectively defined concept that expresses "how well a perceived reaction matches a generated action," and here it is described as the product of the torque, *i.e.* action, and the contact forces, *i.e.* reaction [8,11]. This feedback works as shown in Fig. 3. When a torque is generated to bend the body leftward at a certain body part, that part is pushed to the right. In this case, it is expected that there is a support on the right-hand side, because, otherwise, the body part cannot push itself against the environment effectively. Thus, when bending the body leftward, a contact force from the right makes the reaction well matched with the action (Fig. 3(a)). The feedback works in a manner such that further torque is generated to increase the contact force. Conversely, when a torque generated to bend the body leftward results in a contact force from the left, it is an unexpected event. Thus, the reaction does not match the action (Fig. 3(b)). The feedback works in a manner such that further torque is generated to decrease the contact force.

Decentralized Control for Contraction and Expansion. The natural length of the body trunk masses for ($N_\beta < i \leq N$) is controlled according to the following equation:

$$\bar{l}_{c,i}(t+1) = (1 - \rho_i)l_{c,i-1}(t) + \rho_i l_0 (N_\beta < i \leq N).$$ (13)

Here,

$$\rho_i = \tanh\left\{\kappa_p \left([l_{l,i-\frac{1}{2}} - l_{r,i-\frac{1}{2}}]^2 + [l_{l,i+\frac{1}{2}} - l_{r,i+\frac{1}{2}}]^2\right)\right\},$$ (14)

where κ_p is a positive constant.

The first term on the right-hand side of Eq. (13) originates from the control scheme of earthworm locomotion that we proposed previously [7]. This term works such that a force proportional to the ratio of expansion to contraction is generated, and this makes the bodily wave of contraction propagate from the head to the tail to propel the body forward. The second term on the right-hand side of Eq. (13) works so that the natural length approaches the nominal length l_0. Parameter ρ_i, which determines the contribution of the first and second terms on the right-hand side of Eq. (13), is large at the segments where the body trunk bends, as described in Eq. (14).

Thus, peristaltic motion tends to be generated at the segments where the body trunk is straight. Hence, peristaltic and undulatory motion do not compete with each other, and it is expected that the body can propel itself forward under various environments by using reasonable locomotion patterns.

3   Simulation

Simulations were performed to validate the proposed two-dimensional model. The simulator was written in C language, and the fourth order Runge-Kutta method was used. Parameter values, which were chosen by trial and error,

Table 1. Parameter values employed in the simulations.

Parameter	Value	Dimension
n	3	–
m	0.0327	[kg]
g	9.8	$[\mathrm{ms}^{-2}]$
μ_f	1.58×10^{-7}	$[\mathrm{m}^3]$
μ_n	7.87×10^{-7}	$[\mathrm{m}^3]$
μ_b	7.87×10^{-7}	$[\mathrm{m}^3]$
λ	7371	$[\mathrm{m}^{-1}\mathrm{s}]$
N_α	2	–
N_β	4	–
N	69	–
l_0	0.0174	[m]
n_f	3	–
n_b	1	–
κ_u	19.3	$[\mathrm{m}^{-2}\mathrm{kg}^{-2}\mathrm{s}^4]$
σ	0.0057	[m]
κ_p	7244	$[m^{-2}]$
$\bar{l}_{\max,r}, \bar{l}_{\max,l}$	0.0328	$[m]$
$\bar{l}_{\min,r}, \bar{l}_{\min,l}$	0.0041	$[m]$
$\bar{l}_{\max,c}$	0.0305	$[m]$
$\bar{l}_{\min,c}$	0.0102	$[m]$

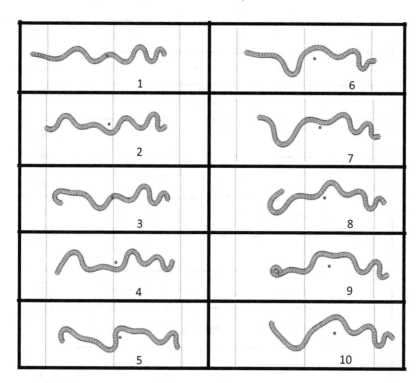

Fig. 4. Simulation result on a flat terrain when $h_u(t)$ is varied with $h_p(t)$ fixed. RTSs generating forces of contraction and expansion are colored blue and red, respectively. (Color figure online)

are shown in Table 1. Higher-level control inputs, $h_u(t)$ and $h_p(t)$, were entered manually via the keyboard by an operator who could get an overview of the simulated robot and the environment from a screen. The operator changed the $h_u(t)$ and $h_p(t)$ values so that the simulated robot can propel itself forward effectively.

First, we performed simulations on a flat terrain, wherein either $h_u(t)$ or $h_p(t)$ was fixed. When $h_u(t)$ was varied with $h_p(t)$ fixed, undulatory locomotion was generated (Fig. 4). When $h_p(t)$ was varied with $h_u(t)$ fixed, peristaltic locomotion was generated (Fig. 5).

Next, we performed a simulation in a narrow aisle (Fig. 6). The simulated robot exhibited concertina locomotion in which the tail part of the body was first pulled forward with the head part anchored followed by extension of the head part with the tail part anchored. Furthermore, peristaltic motion was also used during locomotion, which helped the robot to move effectively.

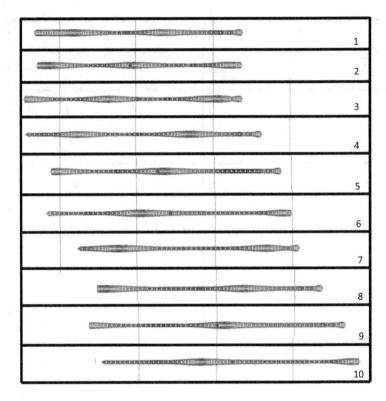

Fig. 5. Simulation result on a flat terrain when $h_p(t)$ is varied with $h_u(t)$ fixed. RTSs generating forces of contraction and expansion are colored blue and red, respectively. (Color figure online)

Finally, we performed a simulation in an extremely narrow aisle that bends at a right angle at two points (Fig. 7). The simulated robot was constrained to move along the aisle. The robot exhibited peristaltic motion for the straight part of the aisle. The robot pushed itself against the wall to move effectively in the areas where the aisle bends.

Thus, the proposed control scheme enabled the simulated robot to move effectively under various environments by using undulatory and peristaltic motion reasonably.

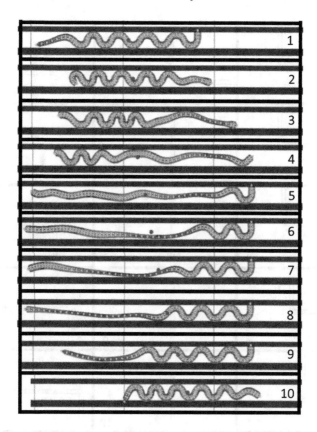

Fig. 6. Simulation result in a narrow aisle. RTSs generating forces of contraction and expansion are colored blue and red, respectively. (Color figure online)

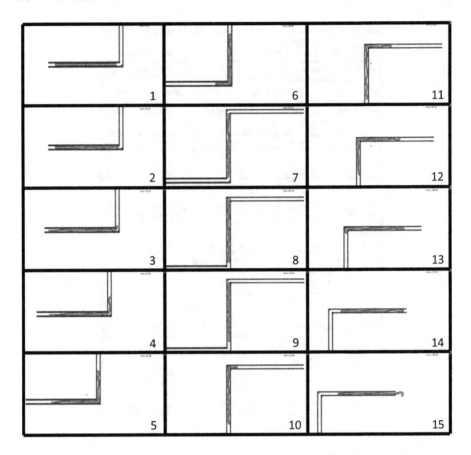

Fig. 7. Simulation result in an extremely narrow aisle that bends at a right angle at two points. RTSs generating forces of contraction and expansion are colored blue and red, respectively. (Color figure online)

4   Conclusion

Our goal was to design a decentralized control scheme that could generate multiple locomotion patterns in response to various environments. As a first step, we proposed a decentralized control scheme that enables well-balanced coupling between undulatory and peristaltic locomotion, by drawing inspiration from snake and earthworm locomotion. We demonstrated via simulation that the robot can exhibit both undulatory and peristaltic locomotion to adapt to various environments in real time.

In the future, we plan to combine legged locomotion with the proposed model. Undulatory, peristaltic, and legged locomotion are the fundamental animal locomotion patterns, and coordination of these three patterns will help achieve highly adaptive locomotion. We also plan to develop a robot and validate the proposed control scheme in the real world.

Acknowledgments. This work was supported by Japan Science and Technology Agency, CREST (JPMJCR14D5), and the Casio Science Promotion Foundation.

References

1. Ijspeert, A.J., Crespi, A., Ryczko, D., Cabelguen, J.M.: From swimming to walking with a salamander robot driven by a spinal cord model. Science **315**, 1416–1420 (2007)
2. Ijspeert, A.J.: Central pattern generators for locomotion control in animals and robots: a review. Neural Netw. **21**, 361–376 (2008)
3. Pfeifer, R., Lungarella, M., Iida, F.: The challenges ahead for bio-inspired 'soft' robotics. Commun. ACM **55**, 76–87 (2012)
4. Liljebäck, P., Pettersen, K.Y., Stavdahl, Ø., Gravdahl, J.T.: Snake Robots - Modelling, Mechatronics, and Control: Advances in Industrial Control. Springer, London (2012). https://doi.org/10.1007/978-1-4471-2996-7
5. Boxerbaum, A., Shaw, K., Chiel, H., Quinn, R.: Continuous wave peristaltic motion in a robot. Int. J. Rob. Res. **31**, 302–318 (2010)
6. Sugita, S., Ogami, K., Michele, G., Hirose, S., Takita, K.: A study on the mechanism and locomotion strategy for new snake-like robot active cord mechanism –slime model 1ACM-S1. J. Robot. Mechatron. **20**, 302–309 (2008)
7. Kano, T., Kobayashi, R., Ishiguro, A.: Decentralized control scheme for adaptive earthworm locomotion using continuum-model-based analysis. Adv. Robot. **28**, 197–202 (2014)
8. Kano, T., Yoshizawa, R., Ishiguro, A.: Tegotae-based decentralised control scheme for autonomous gait transition of snake-like robots. Bioinspiration Biomim. **12**, 046009 (2017)
9. Umedachi, T., Takeda, K., Nakagaki, T., Kobayashi, R., Ishiguro, A.: Fully decentralized control of a soft-bodied robot inspired by true slime mold. Biol. Cybern. **102**, 261–269 (2010)
10. Date, H., Takita, Y.: Adaptive locomotion of a snake like robot based on curvature derivatives. In: Proceedings IEEE/RSJ International Conference on Intelligent Robots and Systems, IROS07, pp. 3554–3559 (2007)
11. Owaki, D., Goda, M., Miyazawa, S., Ishiguro, A.: A minimal model describing hexapedal interlimb coordination: the tegotae-based approach. Front. Neurorob. **11**, 29 (2017)

Calibration Method to Improve Transfer from Simulation to Quadruped Robots

Gabriel Urbain[(⊠)], Alexander Vandesompele, Francis Wyffels,
and Joni Dambre

IDLab, Electronics and Information Systems Department,
Ghent University, imec, Ghent, Belgium
gabriel.urbain@ugent.be

Abstract. Using passive compliance in robotic locomotion has been seen as a cheap and straightforward way of increasing the performance in energy consumption and robustness. However, the control for such systems remains quite challenging when using traditional robotic techniques. The progress in machine learning opens a horizon of new possibilities in this direction but the training methods are generally too long and laborious to be conducted on a real robot platform. On the other hand, learning a control policy in simulation also raises a lot of complication in the transfer. In this paper, we designed a cheap quadruped robot and detail a calibration method to optimize a simulation model in order to facilitate the transfer of parametric motor primitives. We present results validating the transfer of Central Pattern Generators (CPG) learned in simulation to the robot which already give positive insights on the validity of this method.

Keywords: Embodiment · Compliant quadruped locomotion
Transfer learning · CPG

1 Introduction

The progress in *machine learning* within the last decades represents a great opportunity for improving specific performance metrics in the robotics state-of-art [4]. Among possible applications, locomotion constitutes an excellent example as deterministic controllers are not well adapted to unknown terrains and sensitive to external perturbations. Encouraging results involving complex gait control policies using rich sensor information to increase adaptiveness and robustness in an unknown dynamic environment have been introduced in the past years. Among the most influential contributions, *deep reinforcement learning* techniques have shown the most important improvements [12,17]. However, the works are mainly conducted in simulation, whereas the controllers embedded on state-of-the-art robots are still mostly using sophisticated analytic methods with explicit knowledge of the robot physical parameters, e.g. in [2,16,18]. Indeed,

© Springer Nature Switzerland AG 2018
P. Manoonpong et al. (Eds.): SAB 2018, LNAI 10994, pp. 102–113, 2018.
https://doi.org/10.1007/978-3-319-97628-0_9

direct learning on a robot presents several drawbacks, among which the training time –which can be generally sped up in a physics simulation–, the wearing and hysteresis of mechanical parts, but also potential damages when the robot is exploring its own motor control capacities.

Partial training of policies in simulation is a straightforward idea to obtain a good initial state and avoid mechanical damage before transfer to a real robot. This field, known as *transfer learning*, involves big challenges as a lack of accuracy and realism during the simulation can easily lead to a failure in the real environment. However, simply using a more realistic simulator shifts the problem toward a sophisticated implementation of the physics engine and the need of heavier computational resources. Despite efforts in this direction, there is still a huge reality gap with the current simulators. Therefore, transfer in robotics has mainly focused on making controllers robust to changing environments. Among the recent related work, [5] shows how a quadruped locomotion gait, pre-trained in one environment, can learn faster on other surfaces. For higher-level architectures and learning techniques, [6] presents a new neural network architecture for transfer of *reinforcement learning* with various robotic tasks and [13] focuses on locomotion using a hierarchical bio-inspired control architecture combining recurrent networks for motor primitives using proprioceptive feedback together with higher level feed-forward networks also processing visual information.

The *embodiment* theory, and in particular *morphological computation* in its larger sense, i.e. as presented in [8] may be a preliminary solution to theorize the transfer problem. In this framework, body, controller, and their intricate relation are analyzed from a dynamical perspective. Each entity can be modeled as a non-linear filter with computational skills rather than explicitly in a kinematic parameter space. From that point of view, a good simulation works with an accurate representation of the transfer functions rather than a detailed physics implementation. This can be obtained through automated optimization rather than fine-tuning of model parameters. Such an idea has been used for instance with quadruped robots that refine their own body representation by matching real and simulated sensors inputs [3]. With the trends in data-driven learning techniques, an increasing number of studies also consider the simulated physics as a black box with parametric equations, for instance using neural networks [15]. However, this creates a problem with respect to interpretability and the parametric calibration approach presented in this paper using a standard physics engine and a parametric model represents a good compromise.

This paper introduces an automated calibration method for a simulation model that enables optimal transfer of a controller to a mechanical platform. It takes inspiration in *morphological computation*: rather than trying to replicate rigorously the robot physics, we optimize a parametric model to maximize the similarities between simulation and real world of the body sensor-to-actuator transfer function. In the next section, the design and optimization methods are discussed in further detail. The third section shows the results of two experiments: the first concerns the calibration method and the second evaluates the performance of an open-loop controller trained in simulation and transferred on the real robot. Finally, conclusions and perspectives on the next research steps are given in the last section.

2   Methods

2.1   The Robot

Mechanical Design. The platform used in this project is an update of the *Tigrillo* robot [21] shown in Fig. 1. It implements four under-actuated knee joints, whose kinematic constraints can be manually tuned using detachable springs and dampers to tune the passive compliance properties. A special effort was applied to make it low-cost, versatile and reproducible. The robot weighs 950 g and fits in a box of 30 cm by 18 cm. The track widths are 15 cm in the front and 11 cm between the hind legs and the distance between front and back measures 16 cm, providing a stable balance with any configuration of slow gaits. The legs are directly coupled to four *Dynamixel RX-24F* servomotors selected as a compromise between weight, torque and fast rotation speed.

Fig. 1. On the left, the quadruped *Tigrillo* robot used in this paper for experiments on calibration and transfer of control. On the right, the corresponding parametric simulation model in *Gazebo*.

The use of passive compliance in the knees in place of rigid constraints has experimentally shown a decrease of the optimal cost of transportation but also helps to obtain smoother behaviors in the overall locomotion process [21]. The same leg principles have been also presented in research involving robots like Tekken [9], Puppy [1] or Bobcat [14].

Electrical Design. Following the same constraints on reproducibility and cost, the electronics stack is made from three off-the-shelf boards. First, a DC step-up voltage converter supplies the other boards and motors with a 12 V regulated voltage and a stalk current that can rise to 10 A when the legs are pushing together and the motors have to deliver a high torque. Secondly, an *OpenCM* board is used to read the analog sensor values and send the position or velocity commands to the servomotors. The computer board is a *Raspberry Pi 3* running the *Robot Operating System (ROS)*[1] and streaming actuation and sensor signals to a computer over *ROS* topics.

---

[1] http://wiki.ros.org/.

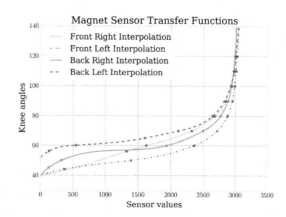

Fig. 2. On the left, a close-up picture of the leg shows the direct coupling with the *Dynamixel* motor and the non-actuated knee joint embedding the angle sensor. On the right, the four sensor calibration functions are represented. Differences in offset and shape of the curves come from the manual fixation of the magnet and the Hall sensor.

Angle Sensors. One important aspect regarding the application of the morphological computation theory is to extract nonlinear feedback from the robot compliant morphology [11]. To this end, permanent rare-earth magnets are attached to the lower parts of each leg and analog *Hall sensors* to the higher parts. The setup presents an advantage in cheapness and does not obstruct the joint movement. This sensor is principally made of a semi-conductor triggered by the Hall effect to output a voltage between $0\,$V and $5\,$V proportionally to the surrounding magnetic field $\boldsymbol{B}$. As a first approximation, this field is decreasing with the cubic value of the distance to the magnet d. This length varies with the square root of the cosine of the knee angle α, according to the generalized Pythagorean theorem. As a consequence, the variation of the sensor voltage is high for small angles and the order of magnitude for the best accuracy is around a hundredth of a degree. It however decreases quickly with the knee angle to reach approximately one degree when the legs are fully extended. For each leg, a conversion table between the sensor value and the measured angle is used to interpolate the transfer function of the sensor presented in Fig. 2. All the curves have a shape conform to the expected aspect but the manual fixation of the magnet and the Hall sensor leads to a different offset for each leg.

2.2   Calibration of the Simulation Model

The approach followed in the calibration method consists in tuning a set of simulation parameters in order to optimally match the simulated body's response to the real robot's response as observed by the sensors, when an actuation signal is applied. The simulation is performed in the *Neurorobotics Platform (NRP)* [7] using *Gazebo* with the *ODE* physics engine. We select the set of parameters $\boldsymbol{\theta}$ both for their importance in the locomotion behavior but also because they are harder to measure or model accurately:

$$\boldsymbol{\theta} = \{k_{\mathrm{f}}, k_{\mathrm{h}}, \mu_{\mathrm{f}}, \mu_{\mathrm{h}}, d_{\mathrm{f}}, d_{\mathrm{h}}, m_{\mathrm{dist}}, \}. \tag{1}$$

In this notation, the index f refers to the front legs of the robot and h refers to the hind legs. The parameters k are the spring constants in the knee joints, expressed in N/m. These parameters are chosen because the complexity of the spring models in the physics engine is rather limited and can lead to insufficient results, as discussed further in the Results section. The parameters d are the contact depth coefficients expressed in m and represent how much two rigid bodies can overlap during simulation to compute the friction forces. They directly interfere with the static friction coefficients μ and a good manual tuning generally requires empirical comparison with the real robot. Finally, the total mass of the robot is fixed and determined by weighting the robot but the distribution ratio between the front and the back is represented with m_{dist}. Many other parameters like the damping values in the knee joints, the minimal value of knee angles, the kinematic friction coefficients of the feet and the motor characteristics parameters have been evaluated in this research but none of them has shown significant improvements with respect to the uncalibrated model. Therefore, they are not discussed further in this paper. Figure 3 shows the architecture of the optimization process used for calibration. The robot is actuated in open-loop with four sinusoidal signals $\boldsymbol{a}(t)$, with the same amplitude and phase for all legs such that its body will alternate between standing-up and sitting-down. The sensor values of the robot $\boldsymbol{s}_{\mathrm{r}}(t)$ are recorded to estimate the robot transfer function:

$$\boldsymbol{S}_{\mathrm{r}}(p) = \mathcal{F}_{\mathrm{r}}(\boldsymbol{A}(p)), \tag{2}$$

where $\boldsymbol{A}(p)$ and $\boldsymbol{S}_{\mathrm{r}}(p)$ are the Laplace transforms of the robot motor and sensor signals $\boldsymbol{a}(t)$ and $\boldsymbol{s}_{\mathrm{r}}(t)$.

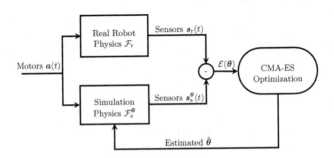

Fig. 3. The calibration is represented on this diagram. Sensor values recorded on the robot are used to optimize the unknown morphology parameters of the simulation model with *CMA-ES*.

An optimization is performed with *covariance matrix adaptation evolutionary strategy (CMA-ES)*, as formulated in [10]. It has the advantage to converge rapidly in a landscape with several local minima and requires only few initialization parameters. The algorithm generates simulation models with different sets of parameters and estimates the one that minimizes the error:

$$\hat{\boldsymbol{\theta}} = \arg\min_{\boldsymbol{\theta}} \ \mathcal{E}(\boldsymbol{\theta}). \tag{3}$$

The error function $\mathcal{E}(\boldsymbol{\theta})$ is chosen to represent the difference between two temporal signals but shall also allow invariance against a possible slight phase shift. This invariance can be obtained by computing the *Mean Absolute Error (MAE)* for different configurations where the robot sensors signal $\boldsymbol{s}_r(t)$ and the simulation sensors signal $\boldsymbol{s}_s(t)$ are shifted within a time window, and finally taking the minimum error:

$$\mathcal{E}(\boldsymbol{\theta}) = \min_{k \in [-M,M]} \ \frac{1}{N} \sum_{i=k}^{N+k} \mathrm{abs}(\boldsymbol{s}_{r,i} - \boldsymbol{s}_{s,i}^{\theta}), \tag{4}$$

where N is the number of samples measured for the two signals during a period T, and M is the boundary of the shifting window. The errors for each sensors are simply added to obtain the final total error value.

Finally, it is assumed that the parameter set that is obtained in the optimization corresponds to a good approximation of the robot transfer function:

$$\mathcal{F}_s^{\hat{\theta}} \approx \mathcal{F}_r. \tag{5}$$

In order to decrease the noise on the sensors, which is caused by different factors like some electro-magnetic perturbations or the undesirable residual vibration coming from the motors, the signal is segmented using the zero-crossing of the periodic actuation signal as a threshold. All the sensors segments are then projected on a one-period domain with a fixed number of points to be easily averaged and compared (see Fig. 4a).

3 Results

3.1 Calibration of the Simulation Model

To apply the calibration method in practice, the robot is actuated by a 0.3 Hz sine wave during one minute and the real motor signals and the sensor signals are recorded during the last 50 s, to provide a reference for the optimization. After this step, a *CMA-ES* algorithm is run on a computer. Different initialization values were tested but a low variance was observed in the results. For each algorithmic epoch, a population of 10 individuals is generated: each of them has a different morphology sampled in the distribution of $\boldsymbol{\theta}$ provided by *CMA-ES* and is simulated with the same actuation pattern as the real robot. The optimization

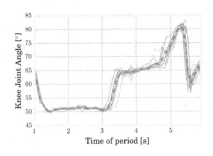

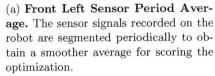

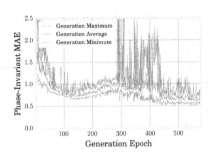

(a) **Front Left Sensor Period Average.** The sensor signals recorded on the robot are segmented periodically to obtain a smoother average for scoring the optimization.

(b) **CMA-ES Evolution.** The evolution of the *CMA-ES* algorithm through the successive generations converges a first time around the 100th generation then move again to find a better minimum at the 500th generation.

Fig. 4. An overview of the optimization method evolution.

evolution presented on Fig. 4b shows a saturation of the *MAE* error to 0.5 after more than 500 generations.

To validate qualitatively this optimum, the *Neurorobotic Patform (NRP)* is used to control together the simulation model and the real robot. A variety of actuation sinusoids with different amplitudes are applied on the motors. A visual comparison indicates a good correlation between the simulation and the real observations[2]. Interestingly, failures like stumbling or falling are also observed simultaneously in both worlds which is very useful as the role of pre-training in simulation is to exclude actuation patterns that lead to instability of the physical robot.

To explain these results, a single sine wave of 2 Hz –the same as during the optimization– is given to the motors. The average sensor signals over one period are plotted for the real robot, the optimum morphology in simulation and the initial default morphology in simulation (see Fig. 5a). The latter model is made with our best knowledge of the robot mechanical parameters and simulator parameters and serves as a baseline. A comparison of the parameters before and after optimization is also given in Table 1. It is quite obvious on the figure that the signals from the optimized morphology are closer to the observation on the robot. This has certainly two major causes. First, the default parameters are unable to reproduce correctly the friction forces between the feet and the ground. This effect depends both on the friction coefficient μ but also the contact depth between two rigid bodies in simulation, which has no concrete meaning in the physical world. During testing, when the robot lifts its body off the ground, the *Center of Gravity (CoG)* move forward above the supporting point of the front feet. The feet suddenly slide on the ground and the knee angle value drops

---

[2] https://youtu.be/CqpkC630fJA.

before to increase again when the robot moves back to the ground and the *CoG* is shifted backwards. This phenomenon is observed on both the real robot and the optimized model but not with the default parameters. Another reason concerns the spring stiffness, which seems too low for the default morphology although it has been fixed by actual measurement on the robot. This is induced by the limitation of the simulator which only allows to model the physical spring via an equivalent torsional spring linear model in the joint. For low spring values, the model is quite accurate but for higher values, the non-linearities induce a saturation on the real robot which cannot be simulated consistently. The optimization copes with this problem by converging on a larger stiffness in the hind legs as displayed in Table 1.

Table 1. This table shows the calibrated morphology parameters before and after optimization. If the mass distribution and the front spring stiffness does not change significantly, the hind knee joints become much stiffer to counteract the spring saturation on the real robot and the friction with the ground decreases in the front and increases in the back to render correctly the general movement.

Morphology parameters	Non-optimized	Optimized
Hind contact depth coefficient (mm)	0.5	7.8
Hind friction coefficient μ	0.1	0.000819
Hind spring stiffness (N/m)	181.6	440.629
Hind mass (kg)	0.238	0.276
Front contact depth coefficient (mm)	0.5	3.8
Front friction coefficient μ	0.1	0.283
Front spring stiffness (N/m)	181.6	181.4691
Front mass (kg)	0.712	0.674
Mean absolute error	3.231	0.483

3.2   Validation with Open-Loop Gaits

In order to corroborate the preliminary observations on the calibration method described in the last section, different open-loop controllers are trained in simulation and we observe how the transfer performs on the real robot.

The controller is modeled by four coordinated *CPGs* with the equations introduced in [19]. Constraints on the frequency and the phases between the different legs are added to obtain motor primitives for walking and bounding gaits at 1 Hz. The other *CPG* parameters are obtained with a *CMA-ES* optimization with the optimized simulation model and where the robot speed is used as a score. This optimization has been successfully conducted in previous research [20] and helps to explore the motor space to find the most stable gaits to locomote along a straight line. The resulting controller does not make use of sensor

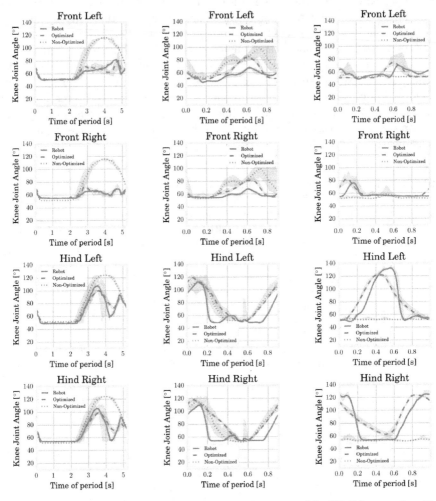

(a) **Optimization**. The signal is closer to the real robot after optimization. This comes from stiffer springs in the hind legs and a different friction that allows more realistic trajectories.

(b) **Bounding gait**. The optimized model is closer to the real robot during the bounding gait. However, it still fails to reproduce correctly the angle variation speed.

(c) **Walking gait**. For the walking gait, the optimized signal gives good results despite a phase shift due to communication delays. However, the non-optimized model directly falls with this gait.

Fig. 5. Sensor signals of the real robot (blue), the optimized simulation model (red) and the default model (green) during optimization, bounding gait and walking gait. (Color figure online)

inputs and only the actuation and the robot mechanical properties are discussed hereafter. Some videos can be also found online[3] to get a better understanding of the locomotion behavior.

Bounding Gait. To come up with a bounding gait, a constraint is added on the *CPGs* phases such that the hind legs are going through stance and swing phases synchronously and the front ones as well. This gait is inherently very stable as it does not involve any movement perpendicular to the walking direction. Figure 5b compares the value of the knee angle on a the real robot as well as the simulation default and optimized models. The optimization introduces better value ranges for the legs but also much more homogeneous results as highlighted by the larger variance of the non-optimized signal. However, it seems that the simulation fail to render the fast transition measured on the robot for both models. In future work, the optimization process should involve higher frequencies to solve this issue.

Walking Gait. These gaits are not especially stable as no efforts were made to optimize the robot's balance in the lateral axis and the robot does not allow feet retraction during locomotion. However, some correct gaits are obtained by optimizing the *CPG* parameters whilst setting the same phase for the legs diagonally opposed. In leads to good performance on the real robot, and the knees angles outlined by the blue curve on Fig. 5c are correctly simulated with the optimized robot model. A phase shift of approximately 100 ms can be explained by the delay introduced by the communication line and the motor inertia. No results could be collected on the non-optimized model for this specific gait, as the robot directly felt on the ground in the simulator. This is not surprising as the controller was optimized using the calibrated model but indicates a good correlation between optimized model and real robot that does not exist with the default model.

4 Conclusions and Future Work

This paper investigated a method to calibrate the simulation model of a cheap passive compliant quadruped robot called *Tigrillo*, in order to transfer efficiently motor primitives learned in simulation.

To derive a usable model of the mechanical platform, it was not chosen to tune them manually but rather to use morphological computation framework to calibrate a parametric simulation model with optimization techniques. From the first observations, this method resulted in a simulation model that can achieve better performance. It also reduces the amount of knowledge and hypothesis that have to be given in the design process. This method has enabled the transfer of parametric *CPGs* for walking and bounding gaits trained in simulation to the

---

[3] https://youtu.be/zCHRWxfoOMU.

real platform while keeping a realistic behavior compared to the observations in the simulated environments.

In further work, this approach should be generalized for a larger range of actuation frequencies to get a more significant optimization score. Also, the transfer could be characterized against other relevant metrics like robot trajectory, pitch or foothold pattern. Finally, the training and transfer of closed-loop motor primitives should also be investigated on the *Tigrillo* platform to enable research on higher-level models.

Acknowledgments. This research has received funding from the European Unions Horizon 2020 Framework Programme for Research and Innovation under the Specific Grant Agreement No. 720270 (Human Brain Project SGA1).

References

1. Aschenbeck, K.S., Kern, N.I., Bachmann, R.J., Quinn, R.D.: Design of a quadruped robot driven by air muscles. In: The First IEEE/RAS-EMBS International Conference on Biomedical Robotics and Biomechatronics, BioRob 2006, pp. 875–880. IEEE (2006)
2. Barasuol, V., Buchli, J., Semini, C., Frigerio, M., de Pieri, E.R., Caldwell, D.G.: A reactive controller framework for quadrupedal locomotion on challenging terrain. In: 2013 IEEE International Conference on Robotics and Automation, ICRA, pp. 2554–2561 (2013)
3. Bongard, J., Zykov, V., Lipson, H.: Resilient machines through continuous self-modeling. Science **314**(5802), 1118–1121 (2006)
4. Connell, J.H., Mahadevan, S.: Robot Learning, vol. 233. Springer, New York (2012). https://doi.org/10.1007/978-1-4615-3184-5
5. Degrave, J., Burm, M., Kindermans, P., Dambre, J., Wyffels, F.: Transfer learning of gaits on a quadrupedal robot. Adapt. Behav. **23**(2), 69–82 (2015)
6. Devin, C., Gupta, A., Darrell, T., Abbeel, P., Levine, S.: Learning modular neural network policies for multi-task and multi-robot transfer. In: 2017 IEEE International Conference on Robotics and Automation, ICRA, pp. 2169–2176 (2017)
7. Falotico, E., et al.: Connecting artificial brains to robots in a comprehensive simulation framework: the neurorobotics platform. Front. Neurorobot. **11**, 2 (2017)
8. Füchslin, R.M., et al.: Morphological computation and morphological control: steps toward a formal theory and applications. Artificial Life **19**(1), 9–34 (2013)
9. Fukuoka, Y., Kimura, H., Hada, Y., Takase, K.: Adaptive dynamic walking of a quadruped robot Tekken on irregular terrain using a neural system model. In: 2003 IEEE International Conference on Robotics and Automation, ICRA, vol. 2, pp. 2037–2042. IEEE (2003)
10. Hansen, N.: The CMA evolution strategy: a comparing review. In: Lozano, J.A., Larrañaga, P., Inza, I., Bengoetxea, E. (eds.) Towards a New Evolutionary Computation. Studies in Fuzziness and Soft Computing, vol. 192, pp. 75–102. Springer, Heidelberg (2006). https://doi.org/10.1007/3-540-32494-1_4
11. Hauser, H., Ijspeert, A.J., Füchslin, R.M., Pfeifer, R., Maass, W.: The role of feedback in morphological computation with compliant bodies. Biol. Cybernet. **106**(10), 595–613 (2012)
12. Heess, N., et al.: Emergence of locomotion behaviours in rich environments. CoRR abs/1707.02286 (2017). http://arxiv.org/abs/1707.02286

13. Heess, N., Wayne, G., Tassa, Y., Lillicrap, T.P., Riedmiller, M.A., Silver, D.: Learning and transfer of modulated locomotor controllers. CoRR abs/1610.05182 (2016). http://arxiv.org/abs/1610.05182

14. Khoramshahi, M., Spröwitz, A., Tuleu, A., Ahmadabadi, M.N., Ijspeert, A.J.: Benefits of an active spine supported bounding locomotion with a small compliant quadruped robot. In: 2013 IEEE International Conference on Robotics and Automation, ICRA, pp. 3329–3334. IEEE (2013)

15. Martius, G., Lampert, C.H.: Extrapolation and learning equations. CoRR abs/1610.02995 (2016). http://arxiv.org/abs/1610.02995

16. Park, H., Wensing, P.M., Kim, S.: High-speed bounding with the MIT cheetah 2: control design and experiments. Int. J. Robot. Res. **36**(2), 167–192 (2017)

17. Peng, X.B., Berseth, G., Yin, K., van de Panne, M.: DeepLoco: dynamic locomotion skills using hierarchical deep reinforcement learning. ACM Trans. Graph. **36**(4), 41:1–41:13 (2017)

18. Raibert, M., Blankespoor, K., Nelson, G., Playter, R.: Bigdog, the rough-terrain quadruped robot. IFAC Proc. Vol. **41**(2), 10822–10825 (2008)

19. Righetti, L., Ijspeert, A.J.: Pattern generators with sensory feedback for the control of quadruped locomotion. In: 2008 IEEE International Conference on Robotics and Automation, ICRA, pp. 819–824 (2008)

20. Urbain, G., Degrave, J., Carette, B., Dambre, J., Wyffels, F.: Morphological properties of mass-spring networks for optimal locomotion learning. Front. Neurorobot. **11**, 16 (2017)

21. Willems, B., Degrave, J., Dambre, J., Wyffels, F.: Quadruped robots benefit from compliant leg designs. Presented at the 2017 IEEE/RSJ International Conference on Intelligent Robots and Systems (2017)

Gait Transition Between Simple and Complex Locomotion in Humanoid Robots

Sidhdharthkumar Vaghani, Yuxiang Pan, Fred Hamker, and John Nassour[✉]

Chemnitz University of Technology, Chemnitz, Germany
{sidhdharthkumar.vaghani,yuxiang.pan}@s2015.tu-chemnitz.de,
{fred.hamker,john.nassour}@informatik.tu-chemnitz.de

Abstract. In this paper, we present the gait transition between rhythmic and non-rhythmic behaviors during walking of a humanoid robot Nao. In biological studies, two kinds of locomotion were observed in cat during walking on a flat terrain and on a ladder (simple and complex walking). Both behaviors were obtained on the robot thanks to the multi-layers multi-patterns central pattern generator model. We generate the rhythmic behavior from the non-rhythmic one based on the frequency of interaction between the robot feet and the ground surface during the complex locomotion. Although the complex locomotion requires a sequence of descending control signals to drive each robot step, the simple one requires only a triggering signal to generate the periodic movement. The overall system behavior fits with the biological findings in cat locomotion.

1 Introduction

Locomotion is an essential type of mobility that allows animals to move in search of food and needs. Different locomotion behaviors were observed in legged animals such as walking, running, and jumping. Neuroscientists study the neural explanation of locomotion [1,2]. The rhythmic movements are explained by the hypothesis of the Central Pattern Generator (CPG) neurons, which is located in the spinal cord of vertebrates [3]. The CPG generates locally rhythmic signals that drive muscle activation during locomotion. However, CPG neurons receive drive signals from other brain areas to modulate the behavior. The concept of the central pattern generator explains through inter-connected neural models the rhythmic activities observed in biological systems. However, it is still a question of debate if discrete movements are generated by similar neural models as those for rhythmic movements [4,5]. Robotic studies have used the concept of the CPG to generate locomotion behaviors in legged robots [6–8]. Behavior transitions remain as

---

Video about this paper: https://youtu.be/-Mrsa0l_na8

© Springer Nature Switzerland AG 2018
P. Manoonpong et al. (Eds.): SAB 2018, LNAI 10994, pp. 114–125, 2018.
https://doi.org/10.1007/978-3-319-97628-0_10

an important research focus in both neuroscience and robotics [9–11]. Gait transitions from walking to swimming based on CPGs have been shown by Ijspeert in his Salamadra robot [11]. The robot switch between two different movements (tasks): walking and swimming. However, behavior transition may occur within the same task. Marlinski et al. recognized two different types of locomotion in the cat while walking [12]. The first is called "simple locomotion", it is observed while the cat walks on a flat regular terrain. The second is called "complex locomotion", it is observed while walking on a horizontal ladder. The study emphasizes the important role of the ventrolateral (VL) thalamus in the cat's brain to contribute to the locomotion related activity. The authors show that the activity of the VL neurons in the cat was more elevated while walking on the ladder (complex terrain) than in flat terrain. This is referred to the supposed role of the VL-thalamus in transmitting signals that control the landing positions into the motor cortex. The discharge of the VL-thalamus neurons is related to the information of the complex environment, some VL neurons discharge in different phases of the cycle during walking on a flat terrain (simple locomotion) and walking on the ladder (complex locomotion) [12]. The study concludes that the VL-thalamus reflects the separation between simple and complex behaviors while walking.

In this paper, we describe the gait transition between the simple and the complex robot locomotion by employing a previously proposed CPG model [8]. The CPG model possesses rhythmic and discrete patterns generated by the same neural model [13]. Thanks to this CPG architecture, simple locomotion is generated by using oscillation patterns and complex locomotion is generated with discrete patterns. Furthermore, we show how a sequence of stimuli that trigger discrete patterns on a robot joint will produce an oscillatory pattern on the same joint. The generated oscillation behaviors will always be present even in the absence of the previous inducing external stimulus. This is an interesting transition from a discrete stimulus-based motion into a rhythmic motion. The advantage of this algorithm in robot locomotion is presented in the walking task. We show how to extract the simple walking locomotion (rhythmic) from the complex one (discrete: step-by-step walking). This paper is organized as follows. Section 2 resumes the CPG model, then it describes a neural network used to modulate the CPG parameters for the amplitude and the direction of the motion. Section 3 describes a transition mechanism of behavior based on an external stimulation signal. Complex and simple locomotion during walking is presented in Sect. 4, then a transition mechanism from complex to simple locomotion is presented in Sect. 4.3. The conclusion is presented in Sect. 5.

2   Pattern Generation

2.1   Central Pattern Generator Model

The Multi-Layers Multi-Pattern Central Pattern Generation (MLMP-CPG) model, proposed in [8], is used to generate the rhythmic and the non-rhythmic behavior of the robot. The CPG for each joint is further divided into three layers: Rhythm generator (RG), Pattern Formation (PF) and Motor Neuron (MN), see Fig. 1. Neurons of these layers are responsible for the generation and the formation of the rhythmic and non-rhythmic motion patterns for a joint by a simple tuning of few parameters.

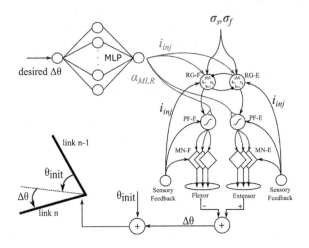

Fig. 1. The MLP neural network modulates the CPG model of each joint by approximating the relationship between (α_{MLR}, i_{inj}) and the variation of joint angle $\Delta\theta$. The MLP input is the desired joint motion $\Delta\theta$, the output is the product of α_{MLR} and i_{inj}.

Rhythm Generation Neurons (RG) are responsible for the generation of various motion patterns like Quiescent, Oscillator and Plateau [13]. The RG neuron model is represented by (1), and (2). A pattern is triggered by injecting a current i_{inj} into the RG neuron.

$$\tau_m.\frac{dV}{dt} = -(fast(V,\sigma_f) + q - i_{inj}), \quad \tau_s.\frac{dq}{dt} = -q + q_\infty(V) \tag{1}$$

$$\tau_m < \tau_s, \quad fast(V,\sigma_f) = V - A_f tanh((\sigma_f/A_f)V) \tag{2}$$

where V is the membrane potential, q is the lumped slow current, $q_\infty(V)$ is the steady-state value of the lumped slow current, τ_m is the membrane time constant, τ_s is the slow current's time constant, i_{inj} is the externally injected current, A_f is the width of the N shape fast IV curve, σ_s and σ_f are the potassium and calcium conductances. The output of the RG neuron is highly related to the parameters σ_f, σ_s, i_{inj} and τ_m.

Pattern Formation Neurons (PF) shape the patterns generated by RG neurons for both flexion and extension [14]. In addition to RG signals, PF neurons receive sensory feedback. The activation function of the PF neuron is given by (3).

$$PF_i = \frac{1}{1 + e^{\alpha.\alpha_{MLR}((\phi+\phi_{MLR})-I)}}, \quad I = \frac{\omega_{rg \to pf}.RG_i + \sum_{j=1}^{n} \omega_j.S_j}{n+1} \quad (3)$$

PF_i is the activation value of the i_{th} pattern formation neuron, $\alpha = 1$ indicates the slope of the Sigmoid function, ϕ is the centre point of the curve that shows the threshold of the neuron, I is the average input to pattern-formation neurons, $\omega_{rg \to pf} = 1$ is the weight between rhythm-generation neurons and pattern-formation neurons, RG_i is the activation of the i_{th} rhythm generator neuron. S_j is the activation of the proprioception or exteroception neuron, ω_j is the weight between this sensory neuron and the pattern-formation neuron (in this paper $\omega_j = 0$). α_{MLR} is a single value that represents the descending control from the high-level controller to modulate the activation of the neuron. By adjusting the value of α_{MLR}, we can control the amplitude of the joint angle. The dominating rhythm (either extension or flexion) is defined by ϕ_{MLR}.

Motor Neurons (MN) perform the flexion and the extension movements of the joint. The activation function for the MN neuron is given by (4).

$$MN_i = \frac{1}{1 + e^{\alpha(\psi-I)}}, \quad I = \frac{\omega_{pf \to mn}.PF_i + \sum_{j=1}^{n} \omega_j.S_j}{n+1} \quad (4)$$

where $\alpha = 1$, $\psi = 0$, I is the input. $\omega_{pf \to mn} = 1$ is the weight between pattern formation neurons and motor neurons (in this paper $\omega_j = 0$).

2.2   Patterns Amplitude Modulation

An approximation function using a multilayer perceptron is used to modulate the amplitude of patterns generated by RG neurons with respect to the desired joint motion. For example, to move the hip roll joint $20[deg]$ from its current configurations we need to adjust α_{MLR} to fit the generated pattern with the desired joint angle. Each CPG is dedicated to one robot's joint. Since each joint has a range for motion that may differ from other joint's range, it is necessary to work out the relationship between the desired joint variation and the modulation parameter separately for each joint. As shown in Fig. 1, the desired joint angle is the sum of the initial angle and the CPG output ($\theta = \theta_{init} + \Delta\theta$). After the motion pattern is completed, the current new joint angle becomes the initial angle. The training data for the neural network are collected by motor babbling. The joint moves randomly based on the motor command provided by the CPG and corresponds to a random α_{MLR}. When the joint reaches the mechanical limitation, the corresponding training data will be eliminated. The neural network for each joint has one input ($\Delta\theta$), one output (α_{MLR}), and one hidden layer with 25 neurons. "ReLu" function was used as an activation function of the hidden

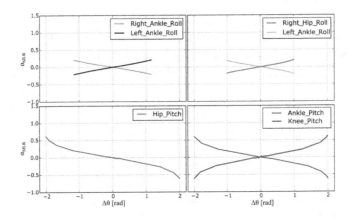

Fig. 2. Joint variations $\Delta\theta$ and α_{MLR} relationships approximated by the MLP.

layer neurons, "Tanh" was used for the output layer. In comparison with the traditional activation function, a "sigmoid", "ReLu" converges faster and can avoid the saturation problem. Mean square error (MSE) was used to calculate the loss, which is the average of the squares of the errors between the predicted value and the target value. We use "Adam" optimizer to minimize the loss. The result of the neural network is shown in the Fig. 2. The average error obtained is around $0.0003[rad]$.

3   Stimulus-Based Behavior Transition

We show how a rhythmic behavior, generated by a CPG, can be constructed from a composition of sequential discrete ones. The advantage of CPG rhythm is that it does not require the presence of feedback, although feedback is important to shape the generated CPG patters. Here, we show how a sensory feedback signal, introduced periodically, is used to trigger a discrete plateau motion on one joint. It leads into a rhythmic movement that can be generated despite the absence of the sensory signal. We carried out this experiment on the left-shoulder-roll joint (LSR). Initially, the joint's CPG was set with the plateau pattern. We used

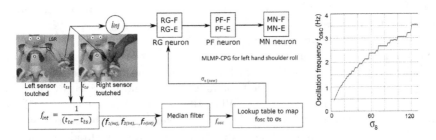

Fig. 3. Stimulus based behavior transition (left). σ_s and the oscillation frequency (f_{osc}) relationship (right).

two touch sensors located on the hand for the stimulus. Every interaction with a touch sensor will trigger a plateau pattern which generates a non-periodic motion for the joint. A periodic interaction with the sensor triggers a sequence of non-periodic plateau patterns, which in turn produces a periodic arm movement. The general block diagram for stimulus-based behavior transition is given in Fig. 3. After each interaction with the touch sensor, the interaction frequency will be measured and filtered to get a smooth signal. The output of the median filter is the oscillation frequency f_{osc} for the respective joint. To generate the periodic movement with the oscillation frequency f_{osc}, the output of the medium filter is mapped to the pre-derived lookup table of the oscillation frequency f_{osc} and the RG neuron parameter σ_s. The relation between the oscillation frequency f_{osc} and the RG neuron parameter σ_s is given in Fig. 3. Once a new σ_s is selected, the oscillatory pattern appears at RG neurons and a rhythmic joint movement starts. Figure 4 shows (1) the periodic and non-periodic movements of the left-shoulder-roll, (2) the interaction frequency, (3) the injected current triggered by the interaction with the touch sensors, and (4) σ_s variation. At the beginning, the joint shows a periodic movement with non-periodic plateau patterns ($\sigma_s = 0.1$). This movement is produced thanks to the rhythmic touch signals of hand's sensors. When an expected activation of touch sensors is delayed (or absent), the CPG switches to an oscillation mode that matches the periodic movement frequency. In the third phase, interaction signals were introduced again to the touch sensors with a higher frequency, the CPG switches back to plateau patterns, a new interaction frequency will be calculated. When the

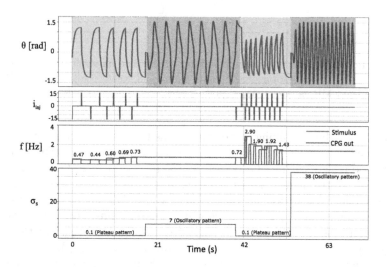

Fig. 4. From top: The periodic and non-periodic joint behaviors of the left-shoulder-roll joint (LSR). The injected current triggered by the interaction with the hand touch sensors. The estimated interaction frequency and the CPG output frequency. σ_s variation based on the interaction frequency in case of osculation patterns and a predefined value ($\sigma_s = 0.1$) in case of plateau patterns.

interaction signals are delayed (or absent) the CPG switches again to an oscillatory mode while matching the last calculated interaction frequency.

4   Complex and Simple Locomotion

4.1   Complex Locomotion

Complex locomotion is referred to the locomotion patterns accompanied by high activation signals in the thalamus of the cat's brain [12]. It was observed during walking on a horizontal ladder. This neural activation was referred to the visual guidance of movement. In this paper, we propose a complex locomotion walking behavior on a Nao humanoid robot. Due to the platform limitations, it is not possible to make the robot walking on a horizontal ladder. Therefore, we have introduced a walking behavior based on discrete CPG patterns. The model acts as a state machine for walking (see Fig. 5(left)).

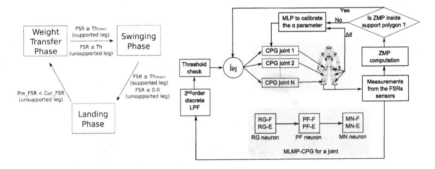

Fig. 5. State machine for step-by-step walking (left). The basic block diagram of the close loop walking locomotion of the NAO humanoid robot using the MLMP-CPG model with the ZMP as a feedback signals (right).

The non-rhythmic walking movement means that each walking step may differ from the previous one. In the complex locomotion after each walking step, the robot should trigger each CPG to perform the next step. The state diagram has mainly three states namely weight transfer phase, swinging phase, and landing phase. The transition between the states is based on the value of the FSR sensors mounted in the robot's feet. In the **weight transfer phase**, the robot transfers his weight on one leg to free the other leg to swing. At the end of this state, the robot will be in a single support phase. In the **swinging phase**, the robot moves one leg forward to perform a step. In the **landing phase**, two movements are performed by the robot. One movement is extending the knee joint and the other is transferring back the weight to the front side, and the robot will be in the double support phase. At this stage, we get the interaction frequency which represents how many times the robot steps ahead each second. The complex walking locomotion is generated using plateau patterns at each joint with the

CPG parameters configured as follows. For rhythm generation neurons $\sigma_s = 0, 1$, $\sigma_f = 5$, $i_{inj} = 1$, and $\tau_m = 0.6$. The parameters of pattern formations varies to provide different ranges of movement at each joint ($\alpha_{MLR} = 0.01$ for hip-roll, $\alpha_{MLR} = 0.03$ for hip-pitch, $\alpha_{MLR} = 0.16$ for ankle-roll, $\alpha_{MLR} = 0.03$ for ankle-pitch, and $\alpha_{MLR} = 0.2$ for knee-pitch). To overcome the stability problem we used the concept of the zero moment point (ZMP) [15]. ZMP is used to compensate the pattern formation parameters at the ankle joint roll and pitch (α_{MLR}). ZMP can be calculated in two ways, the first uses the kinematic model of the robot while the second uses only the information about the center of pressure (using Nao robot FSRs). In this paper, we have selected the second option as a model-free ZMP computation.

$$X_{ZMP} = \frac{\sum_{j=1}^{n} f_i x_i}{\sum_{j=1}^{n} f_i}, Y_{ZMP} = \frac{\sum_{j=1}^{n} f_i y_i}{\sum_{j=1}^{n} f_i} \tag{5}$$

Here, X_{ZMP}, Y_{ZMP} are the X-Y position of the ZMP in the supported polygon. f_i is the measured value by the i_{th} FRS sensor and x_i, y_i are the positions of the i_{th} FSR sensors with reference to the ankle frame of the NAO robot. If the location of the ZMP is inside the support polygon the robot is in balance and there is no need to tune the α_{MLR} parameters in the next walking step. When the ZMP lies outside or on the boundary of the support polygon, the α_{MLR} of the pattern formation neuron of the ankle joints will be updated accordingly. Joint trajectories, which are generated by the MLMP-CPG model, along with the injected current i_{inj} are shown in Fig. 6. The position of the ZMP in the supported polygon for several robot steps is also presented. These reference lines represent the safe-margins from the boundaries of the supported polygon.

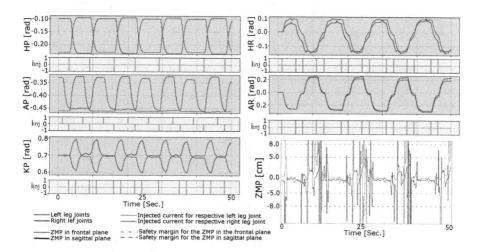

Fig. 6. Joint trajectories in complex walking locomotion with ZMP compensation. The ZMP position represented in the frontal and sagittal planes.

4.2   Simple Locomotion

Simple locomotion is referred to the locomotion patterns, which are non-accompanied by high activation in the thalamus of cat's brain [12]. It was observed during walking on a flat terrain. In this paper, we propose a simple locomotion walking on the Nao humanoid robot based on the complex walking. To define the oscillation parameters and the phase difference between roll and pitch motors, we have employed the previously proposed mechanisms in Sect. 3 to capture the interaction frequency of the complex locomotion. For the experiment of a simple locomotion, the frequency of interaction has been extracted from the complex locomotion. In complex locomotion, we measured the time for each interaction between the robot foot and the ground surface. This timing analysis of FSR sensory signals gives us the frequency of interaction with the ground. The interaction frequency from the complex locomotion is adopted in the simple locomotion by setting the RG neuron parameter σ_s and σ_f. As shown in Fig. 6 the interaction frequency of the robot was very small due to the state machine conditions, it is approximately $0.14[Hz]$, which was adopted by the simple locomotion. To increase the walking speed, we have varied the oscillation frequency by changing σ_s and σ_f. We tested our model in simulation (V-rep) and on the Nao real robot. Various walking speeds have been obtained in the simple locomotion mode. Figure 7 shows the robot joints angle during slow and fast walking speeds. We achieved in the fast speed nearly 1 step per second. The step length throughout the locomotion is approximately $9[cm]$. It can be identified from the phase diagram for both slow and fast walking speeds that the limit cycle for torso angle λ with the vertical and its derivative has a

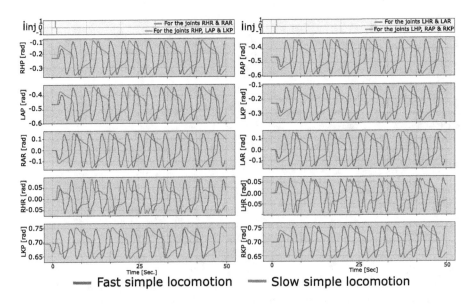

Fig. 7. Joint trajectories during the simple walking with two different speeds.

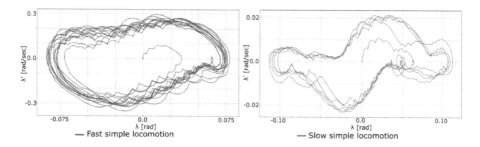

Fig. 8. The phase diagram for the simple locomotion for two different walking speeds.

fix orbit, which proves the stability, see Fig. 8. In the next section, we explain the transition from the simple to the complex locomotion.

4.3 Gait Transition for Adaptive Behavior

As we discussed in the previous section, experiments for the complex locomotion and the simple locomotion were carried out separately. Figure 9 shows the block diagram for the gait transition between simple and complex locomotion. As shown in Fig. 10, the robot walks with the simple locomotion mode at the beginning and after few walking steps it moves to the complex mode then it ends with the simple one. There is a transition pattern (in yellow) in between two consequence locomotion modes. The transition pattern moves the robot to the next state from which the next locomotion mode starts. The transition pattern consists of a plateau pattern generated by the CPG. It must guarantee a smooth transition, otherwise the robot changes suddenly from one locomotion mode into another which may cause falling. When the robot walks with the complex locomotion behavior, the interaction frequency is continuously measured at each interaction between the robot foot and the ground surface. Once the transition occurs, this interaction frequency is mapped to the lookup table of the σ_s and the oscillation frequency. The output of this mapping is the σ_s for the simple locomotion. With this newly derived σ_s, the robot starts to walk with the simple locomotion. Note that the oscillation frequency for the simple locomotion is nearly as same as the interaction frequency in the complex locomotion. Once the simple locomotion starts, the movement will be rhythmic. In this way, the gait transition from the complex to simple locomotion has been performed. It is important to note that the interaction frequency only plays a role when the

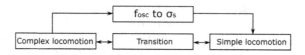

Fig. 9. Block diagram of gait transition between complex and simple locomotion.

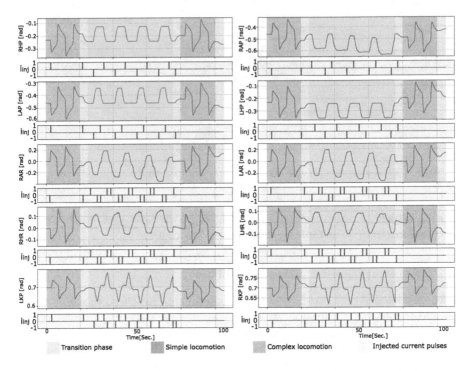

Fig. 10. Joint trajectories during gait transition. (Color figure online)

transition from the complex locomotion to the simple one occurs. For the transition from simple locomotion to complex locomotion, the robot only cares about the current position of all the necessary joints to start the complex mode.

5   Conclusion

The goal of this paper is to present gait transitions from simple to complex locomotion and vice-versa on a humanoid robot while walking. This study proposes one single framework that can host both modes of locomotion. It shows how a periodic pattern can be extracted from a non-periodic one based on the interactions between the robot and the environment. It can be analysed that during the complex walking mode (a sequence of non-rhythmic neural activations) there are more modulation signals of the CPG from the high-level controller as compared to the simple walking mode (rhythmic neural activation). These results are supported by a biological study in the cat locomotion [12]. The complex locomotion requires a sequence of activations from the high-level controller, whereas the simple locomotion requires only one-time activation to generate the rhythmic movement. The complex walking mode has the advantage to modulate each robot step differently as a function of the environment complexity (walking on a ladder in the cat experiment). Due to Nao robot limitation in shorter steps length compared to the feet size, it is not possible to perform the ladder experiment on this robot. Our future work should aim to involve visual feedback to modulate each step while walking in a complex environment, e.g. using a robotic platform with smaller feet size.

References

1. Shik, M.L., Orlovsky, G.N.: Neurophysiology of locomotor automatism. Physiol. Rev. **56**(3), 465–501 (1976)
2. Whelan, P.J.: Control of locomotion in the decerebrate cat. Prog. Neurobiol. **49**(5), 481–515 (1996)
3. Graham-Brown, T.: The intrinsic factors in the act of progression in the mammal. Proc. Royal Soc. B Biol. Sci. **84**(572), 308–319 (1911)
4. Schaal, S., Sternad, D., Osu, R., Kawato, M.: Rhythmic arm movement is not discrete. Nature Neurosci. **7**(10), 1136–1143 (2004)
5. Lefevre, P., Ronsse, R., Sternad, D.: A computational model for rhythmic and discrete movements in uni- and bimanual coordination. Neural Comput. **21**(5), 1335–1370 (2009)
6. Manoonpong, P., Geng, T., Kulvicius, T., Porr, B., Wörgötter, F.: Adaptive, fast walking in a biped robot under neuronal control and learning. PLOS Comput. Biol. **3**(7), 1–16 (2007)
7. Ijspeert, A.J.: Central pattern generators for locomotion control in animals and robots: a review. Neural Netw. **21**(4), 642–653 (2008)
8. Nassour, J., Hénaff, P., Ben Ouezdou, F., Cheng, G.: A study of adaptive locomotive behaviors of a biped robot: patterns generation and classification. In: Doncieux, S., Girard, B., Guillot, A., Hallam, J., Meyer, J.-A., Mouret, J.-B. (eds.) SAB 2010. LNCS, vol. 6226, pp. 313–324. Springer, Heidelberg (2010). https://doi.org/10.1007/978-3-642-15193-4_30
9. Owaki, D., Ishiguro, A.: A quadruped robot exhibiting spontaneous gait transitions from walking to trotting to galloping. Sci. Rep. **7** (2017). Article no. 277
10. Danner, S.M., Shevtsova, N.A., Frigon, A., Rybak, I.A.: Computational modeling of spinal circuits controlling limb coordination and gaits in quadrupeds. eLife **6**, e31050 (2017)
11. Ijspeert, A.J., Crespi, A., Ryczko, D., Cabelguen, J.M.: From swimming to walking with a salamander robot driven by a spinal cord model. Science **315**(5817), 1416–1420 (2007)
12. Marlinski, V., Nilaweera, W.U., Zelenin, P.V., Sirota, M.G., Beloozerova, I.N.: Signals from the ventrolateral thalamus to the motor cortex during locomotion. J. Neuro Physiol. **107**(1), 455–472 (2011)
13. Rowat, P.F., Selverston, A.I.: Learning algorithms for oscillatory networks with gap junctions and membrane currents. Network **2**(1), 17–41 (1991)
14. McCrea, D.A., Rybak, I.A.: Organization of mammalian locomotor rhythm and pattern generation. Brain Res. Rev. **57**(1), 134–146 (2008)
15. Vukobratovic, M., Borovac, B.: Zero-moment point – thirty five years of its life. Int. J. Humanoid Rob. **1**(1), 157–173 (2004)

Detecting a Sphere Object with an Array of Magnetic Sensors

Byungmun Kang$^{(\boxtimes)}$ and DaeEun Kim$^{(\boxtimes)}$

Biological Cybernetics Lab, School of Electrical and Electronic Engineering,
Yonsei University, 50 Yonsei-ro, Seodaemun-gu, Seoul 120-749, Korea
{kbmang,daeeun}@yonsei.ac.kr
http://cog.yonsei.ac.kr/

Abstract. In this paper, we developed an inventive magnetic field sensing system using an array of magnetic sensors inspired by animals using a huge natural Earth's magnetic filed such as pigeons and red foxes. Inspired by their biological behaviors and the animals' compassion-like sensory nerves, a magnetic field sensing system was constructed using eight magnetic sensors. This is not a simple combination of sensors, but rather a circular arrangement of eight analog magnetic sensors with a DC solenoid in the middle. So, this system can measure the signal from both the solenoid's signal and the surrounding magnetic object simultaneously. Through this functions, we can find a global direction of the sensing system and estimate the direction of the magnetic object around the system. We also used a method of locating a metal sphere, which was inspired by the way a sand scorpion estimates its prey. Applying this to the system, it is the purpose of this paper to estimate the direction of its objects using changes in magnetic fields caused by magnetic object.

Keywords: Magnetic field · Analog magnetic sensor · Biomimetics
Sand scorpions · Pigeons · Red foxes

1 Introduction

All living creatures on Earth live using natural signals. Various natural signals, such as light, vibration, magnetic field, and sound, play an important role in living things. For example, they use natural signals in various situations, such as estimating the location of their prey, avoiding enemies, and communicating with each other. In this study, we focus on living creatures by using the largest natural signal, the Earth's magnetic field. There are a variety of animals with this huge signal to apply to their survival strategies. A typical animal using this huge natural signal is a pigeon [1–3]. Pigeons use the direction of the Earths magnetic field to navigate by determining the direction they are heading or their final destination. One other animal that uses the Earth's magnetic field is the red fox. When red foxes find prey, such as a mouse in the snow, they align their bodies to the north and sense the Earth's magnetic field when they jump their

© Springer Nature Switzerland AG 2018
P. Manoonpong et al. (Eds.): SAB 2018, LNAI 10994, pp. 126–135, 2018.
https://doi.org/10.1007/978-3-319-97628-0_11

prey (this is called "mousing") [4,5]. Thus, various animals, including insects [6,7], also apply magnetic fields to their survival strategies. Many biologists and engineers have been interested in how they can detect the Earth's magnetic field. As a result, engineers have developed analog magnetic sensors that are being used in various industries today.

Here, we introduce a new and inventive magnetic field sensing system using an array of analog magnetic sensor. This is not a simple combination of sensors but a circular structure for detecting the surrounding magnetic objects in all directions of the system. The system can also detect the magnetic field's strength of any nearby magnetic object and measure the Earth's magnetic field, a large natural signal. Also, we could get two separate signals by applying a solenoid to the system. One is the typical value of magnetic field strength that can be obtained from an analogue magnetic field sensor and the other can be obtained the amplitude value of the magnetic field strength produced and varied by the solenoid. Therefore, we defined the typical magnetic field sensor value as a DC value and the sensor value produced by the solenoid as an AC value. These two sensor values can be applied to the magnetic field sensing system for estimating the global direction as well as an algorithm for estimating the direction of a surrounding magnetic object. The two sensor values are used to localize the magnetic objects around the system and to estimate the global direction of the system. The method of estimating the direction of the magnetic object is inspired by the method of estimating the direction of the prey in a sand scorpion [8]. The method of estimating the prey by using the vibration signal of the desert scorpion is applied to estimate the direction of the magnetic object by using the sensor values of the sensing system, which are changed by the surrounding magnetic objects. We also demonstrated this through experiments.

2   Method and Experiments

2.1   Construction of the Magnetic Field Sensing System

The magnetic field sensing system introduced in this paper basically consists of eight analog sensors and a solenoid, as shown in Fig. 1. The analog sensor used in this system is a three-axis micro magnetic field sensor module (P0-MAA-01) attached to Aichsteel's AMI 302 chip. This sensor can selectively sense the x-, y-, and z-axes through pin control rollers of CS, CH1, and CH2. In addition, the sensing range of this sensor can be measured up to $\pm 0.2\,$mT, and the sensitivity is $\pm 2.4\,$mV/uT. The system was constructed by placing eight of these analogue magnetic field sensors in a circular 45° interval in this system.

The solenoid located in the center of this magnetic field sensing system is a 12-V DC solenoid. It differs from other magnetic field sensing systems in that it applies a 5-V square pulse signal to the solenoid through an Arduino. At this time, a solenoid circuit for securing current was added.

As mentioned, we can obtain two signals from this system. We defined and used these signals as DC and AC signals. The DC signal means the change in voltage that can be obtained from the magnetic field sensing system when the

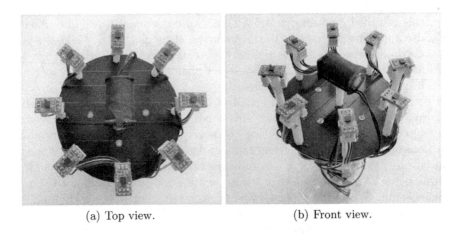

(a) Top view.   (b) Front view.

Fig. 1. The magnetic field sensing system.

solenoid is switched off. The AC signal refers to the amplitude of the voltage change from the magnetic field sensing system when the solenoid is operating. The solenoid produces a 1 Hz square pulse signal which is detected by the sensing system. Then, this signal is defined as the AC value by which the amplitude value of the peak to peak is varied by the magnetic object. In the case of the DC signal, the offset value of analog magnetic field sensor is changed by the magnetic field of the magnetic object as the DC value. In the case of AC signal, when the solenoid is activated, there is a voltage change that can be obtained from the magnetic field sensing system, which is determined by the nearby magnetic object to reduce this variation. This phenomenon was confirmed through experiments.

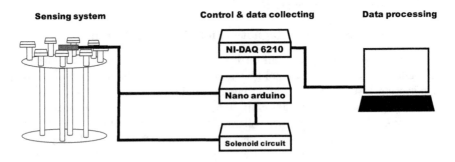

Fig. 2. The diagram of magnetic field sensing system.

Thus, when the system detects a magnetic object around it, it periodically switches the solenoid off and on to obtain this signal and apply it to the direction estimation algorithm. The actual system is constructed as shown in the Fig. 1.

Also, as shown in Fig. 2, it is composed of the NI-DAQ6210, which is used for data acquisition; the Arduino, which is used for pin control of the sensor; and a computer for analyzing data. The system was used to conduct an experiment to estimate the direction of the magnetic object around the magnetic field sensing system.

2.2   Experiments

Experiments were conducted by placing a metal sphere with a diameter of 3 cm around the magnetic field sensing system according to five degrees. Also, the metal sphere was placed 5 cm away from the sensing system. The magnetic field strength measured by the metal sphere in this system can be measured up to a distance of 15 cm, which is set to 5 cm because it is the best measured point of 3 to 5 cm. The height of the metal sphere was set as 15 cm which is same as the height of the system. The experimental environment is as shown in Fig. 3.

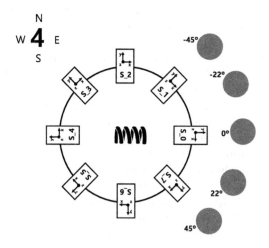

Fig. 3. The environment of experiment.

The five degrees of the metal sphere were 45°, 22°, 0°, −22° and −45° based on the center sensor. Experimental data were also obtained by repeating more than five times of experiments at each degree. In addition, the direction in which the experiment proceeded was in the east direction. The change of the DC value and the change of AC value were measured and applied to the direction estimation algorithm.

2.3   Method of Estimating the Direction of the Magnetic Object

The algorithm for estimating the direction of the magnetic object used here imitates the method of estimating the prey location of a sand scorpion. The sand scorpion estimates the direction of the prey through the time difference of the vibration signal transmitted to the vibrating sensory nerve of its legs when estimating the prey while surviving in a desert environment. This method can estimate the degree even when the vibration transmitted through the medium is sensed through the sensory nerve and the propagation speed of the vibration is unknown. Through a following equation, the direction is estimated by setting the score using the time difference between the vibrations in sand scorpion [9].

$$ze^{i\phi} = \sum_{k=1}^{m} z_k e^{i\phi_k} = x + yi \tag{1}$$

$$x = \sum_{k=1}^{m} z_k \times cos(\phi_k), \quad y = \sum_{k=1}^{m} z_k \times sin(\phi_k) \tag{2}$$

$$\phi = arctan(y/x) \tag{3}$$

ϕ is the direction of the source of the vibrations, and the vibration score for each leg of the sand scorpion is z_k (m = 8), and the degree of the legs is ϕ_k, which can be expressed as above Equations [9]. In Eq. (1), $z_k e^{i\phi}$ can be expressed as a complex number multiplied by an imaginary number with cosine and sine values through Euler formula. Then, according to the score and degree of each leg, it can be expressed as Eq. (2). The values of X and Y can be obtained through Eq. (2) and estimated the direction of the vibration's source (ϕ) through Eq. (3) [9,10]. Therefore, instead of the arrival time of the vibration signal, the DC signal which is the changes of the magnetic field strength and the AC signal which is the changes by the metal sphere and the solenoid measured by the magnetic field sensing system were applied to the above equations.

Therefore, imitating how sand scorpion can estimate the direction of its prey, we used the same method to estimate the direction of the magnetic object. The algorithm used in this paper is to give a score to each sensor in the order of the magnitude of change of the DC value and AC value.

There are three ways to set the score. The first method is to set the score using only the changes in DC value. The second method is to set the score using only the changes in AC value. Finally, the third method is to use the DC values and AC values to set the score, and the results were derived.

3   Results

In this section, the experimental results obtained using the magnetic field sensing system introduced previously are presented. The results are presented in three groups. First is the result of the global direction estimation of this system through the tendency of the variation of DC offset when the magnetic field sensing system is rotated, and second is the variation of DC and AC values for the metal sphere around the system. Finally, we applied the sensor value to the direction estimation algorithm to derive the result for estimating the direction of the magnetic object which is metal sphere.

3.1 Using DC Signal for Global Direction

This magnetic field sensing system can estimate the global direction of the sensing system through the variation of the DC offset value. Figure 4 shows the variation of x-, y- and z-axis of each sensor when the system is rotated in place. In the case of the z-axis, there is no change in value, because the experiment was conducted at the same position and height of the system.

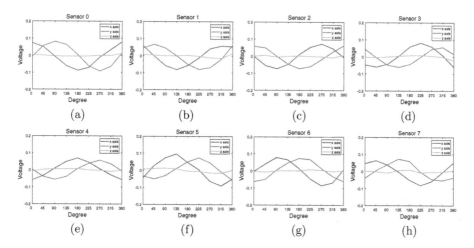

Fig. 4. The DC value of each sensor during rotating. For each sensor, it has a particular x-axis and y-axis value at a particular degree. For the z-axis, the values are constant because the magnetic field sensing system rotates in place with the same height, and data are obtained. (a): sensor 0, (b): sensor 1, (c): sensor 2, (d): sensor 3, (e): sensor 4, (f): sensor 5, (g): sensor 6, (h): sensor 7.

Through this tendency, all sensors can have specific values according to a specific degree. This system can estimate the global direction of the sensing system using this feature.

3.2 Change of the DC Value and AC Value According to the Position of the Metal Sphere in Each Sensor

Figure 5 shows the result of the change in the DC value of the magnetic field sensing system with respect to the five degrees of the metal sphere. This is the value of the change in the x-axis, y-axis, and z-axis of each sensor with $\sqrt{\delta X^2 + \delta Y^2 + \delta Z^2}$. It can be seen that the DC variation of sensor near to each degree is large.

Figure 6 shows the result of the method shown above in Fig. 5, which shows the variation of the AC value measured by the sensing system, when the solenoid is activated. The values of DC and AC can be seen to produce the largest value of magnetic field variation at the sensor where the metal sphere is closest to

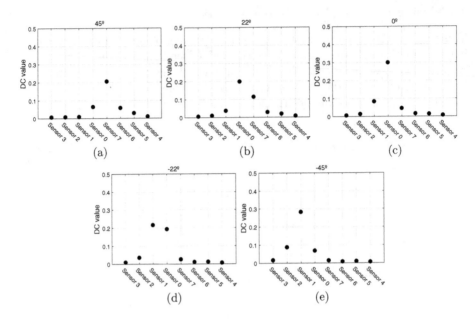

Fig. 5. The DC variation of the each sensor in the sensing system depending on the degree (45°, 22°, 0°, −22°, −45°)

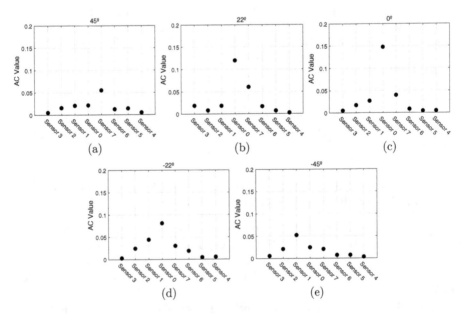

Fig. 6. The AC variation of the each sensor in the sensing system depending on the degree (45°, 22°, 0°, −22°, −45°)

each position. This provided enough information to estimate the location of the metal sphere.

3.3   Estimating the Direction of the Magnetic Object

This result is applied to the method of estimating prey location in the sand scorpion mentioned in Sect. 2.3.

In Fig. 7(a), the direction is estimated for five degrees. The results of the three methods are given by applying only the DC value, only the AC value, and the AC and DC values to the scores of Eqs. (1), (2), and (3). The diagonal blue line is the value of the target degree, the black triangle is the result of DC value, the red circle is the AC value and the blue square is the result of using the AC and DC values. The average error degrees of each are 3.7049° 14.9156° and 1.6807°. In this way, it is more accurate to set the score using both AC and DC values.

Figure 7(b) shows the result of error degree at each degree. It can be shown that when both the AC and the DC values are used, it is possible to estimate the direction surrounding the metal sphere.

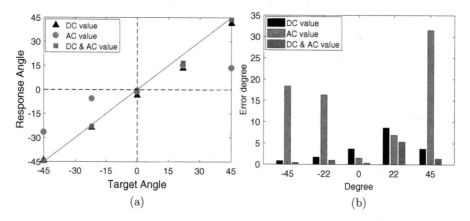

Fig. 7. The three method of estimating the direction of the magnetic sensor (Color figure online)

4   Conclusion and Future Works

In this paper, we introduced a magnetic sensing system that was originally developed using magnetic sensors inspired by living animals by applying a huge natural signal, the Earth's magnetic field, to survival strategies. We also applied the sensor value obtained from this system to the prey location algorithm from sand scorpions' survival strategy to estimate the direction of the magnetic object. In addition, the global direction of the system is estimated from the DC offset changes in the sensor values. The results showed that, although the system was not entirely correct for the object, it could estimate a direction with an average error degree of about 1.68°. In addition, trends in specific sensor values at a specific degree made it possible to recognize the global direction of the system. In addition, the results of this paper only test the metal sphere at a fixed degree, which can be obtained by moving data in real time. It is also possible to estimate the direction of moving objects.

We are currently experimenting with various magnetic objects, and have experimentally proven that they are now also reacting to weak strength of magnetic material such as Aluminum and Nickel. Through the experiments, we plan to develop a magnetic field sensing system to detect the characteristics of magnetic objects in the future.

The above results are important for estimating the direction of the magnetic object. Currently, they are the results of experimental data for five degrees, but it can be a more accurate and robust magnetic sensing system if experiments are conducted at various degrees, even in various distances as well. In addition, the system configuration and the estimation method for a magnetic object have a great advantage in that they are implemented based on biomimetics. This magnetic system also applies to vision-based navigation [11], and it can also be used for robust navigation techniques through interaction in the interior space of a specific magnetic object.

Acknowledgment. This work was supported by the National Research Foundation of Korea (NRF) grant funded by the Korea government (MSIT) (No. 2017R1A2B4011455).

References

1. Fleissner, G., Stahl, B., Thalau, P., Falkenberg, G.: A novel concept of Fe-mineral-based magnetoreception: histological and physicochemical data from the upper beak of homing pigeons. Naturwissenschaften **94**, 631–642 (2007)
2. Davila, A.F., Fleissner, G., Winklhofer, M., Petersen, N.: A new model for a magnetoreceptor in homing pigeons based on interacting clusters of superparamagnetic magnetite. Phys. Chem. Earth Parts A/B/C **28**, 647–652 (2003)
3. Dickman, J.D.: Spatial orientation of semicircular canals and afferent sensitivity vectors in pigeons. Exp. Brain Res. **111**(1), 8–20 (1996)
4. Červený, J., Begall, S., Koubek, P., Nováková, P., Burda, H.: Directional preference may enhance hunting accuracy in foraging foxes. Biol. Lett. **7**(3), 355–357 (2011). rsbl20101145
5. Wiltschko, R.: Magnetic Orientation in Animals. Springer, Heidelberg (2012). https://doi.org/10.1007/978-3-642-79749-1
6. Blakemore, R.P., Frankel, R.B., Kalmijn, Ad.J.: South-seeking magnetotactic bacteria in the Southern Hemisphere. Physics 159 (1980)
7. Newland, P.L., Hunt, E., Sharkh, S.M., Hama, N., Takahata, M., Jackson, C.W.: Static electric field detection and behavioural avoidance in cockroaches. J. Exp. Biol. **211**(23), 3682–3690 (2008)

8. Brownell, P.H., Leo van Hemmen, J.: Vibration sensitivity and a computational theory for prey-localizing behavior in sand scorpions. Am. Zool. **41**(5), 1229–1240 (2001)
9. Kim, D.: Neural network mechanism for the orientation behavior of sand scorpions towards prey. IEEE Trans. Neural Netw. **17**(4), 1070 (2006)
10. Adams, S.V., Wennekers, T., Bugmann, G., Denham, S., Culverhouse, P.F.: Application of arachnid prey localisation theory for a robot sensorimotor controller. Neurocomputing **74**(17), 3335–3342 (2011)
11. Lee, C., Yu, S.-E., Kim, D.E.: Landmark-based homing navigation using omnidirectional depth information. Sensors **17**(8), 1928 (2017)

Cylindrical Terrain Classification Using a Compliant Robot Foot with a Flexible Tactile-Array Sensor for Legged Robots

Pongsiri Borijindakul[1,2,3]($\boxtimes$), Noparit Jinuntuya[1], Alin Drimus[4],
and Poramate Manoonpong[2,3]

[1] Kasetsart University, Bangkok, Thailand
prince_jai-yen@hotmail.com, fscinpr@ku.ac.th
[2] Embodied AI & Neurorobotics Lab, Centre for Biorobotics, The Mærsk Mc-Kinney
Møller Institute, University of Southern Denmark, Odense, Denmark
poma@mmmi.sdu.dk
[3] Institute of Bio-inspired Structure and Surface Engineering,
Nanjing University of Aeronautics and Astronautics, Nanjing, China
[4] The Mads Clausen Institute, University of Southern Denmark,
Sønderborg, Denmark
drimus@mci.sdu.dk

Abstract. In this paper, we present a new approach that uses a combination of a compliant robot foot with a flexible tactile-array sensor to classify different types of cylindrical terrains. The foot and sensor were installed on a robot leg. Due to their compliance and flexibility, they can passively adapt their shape to the terrains and simultaneously provide pressure feedback during walking. We applied two different methods, which are average and maximum value methods, to classify the terrains based on the feedback information. To test the approach, We performed two experimental conditions which are (1) different diameters and different materials and (2) different materials with the same cylindrical diameter. In total, we use here eleven cylindrical terrains with different diameters and materials (i.e., a 8.2-cm diameter PVC cylinder, a 7.5-cm diameter PVC cylinder, a 5.5-cm diameter PVC cylinder, a 4.4-cm diameter PVC cylinder, a 7.5-cm diameter hard paper cylinder, a 7.4-cm diameter hard paper cylinder, a 5.5-cm diameter hard paper cylinder, a 20-cm diameter sponge cylinder, a 15-cm diameter sponge cylinder, a 7.5-cm diameter sponge cylinder, and a 5.5-cm diameter sponge cylinder). The experimental results show that we can successfully classify all terrains for the maximum value method. This approach can be applied to allow a legged robot to not only walk on cylindrical terrains but also recognize the terrain feature. It thereby extends the operational range the robot towards cylinder/pipeline inspection.

Keywords: Compliant robot foot · Flexible tactile-array sensor Cylindrical terrains

© Springer Nature Switzerland AG 2018
P. Manoonpong et al. (Eds.): SAB 2018, LNAI 10994, pp. 136–146, 2018.
https://doi.org/10.1007/978-3-319-97628-0_12

1   Introduction

Currently, walking robots are widely employed for locomotion on complex terrains as well as terrain classification [1,2]. While classifying flat and rough terrains are typical ones for the robots [3–9], classifying cylindrical terrains are still under investigation. Tactile sensing is one of the core sensing methods and the most useful technique for object exploration and terrain recognition [10,11]. It has been widely used in robot hands to classify different shapes, materials, and surfaces [12–14]. Therefore, in this work, we apply a flexible tactile-array sensor to a robot leg with a bio-inspired compliant foot to classify different types of cylindrical terrains during walking. There is a variety of cylinder terrains, for instance, water pipes, gas pipes, wires, and so on. If a robot can classify or recognize the terrains during moving or walking on them, it would be useful for adaptation to the terrains as well as terrain/object inspection. In the following section, we present the robot leg with the compliant foot. Section 3 provides neural locomotion control of the leg for walking on different cylindrical terrains. Section 4 introduces the flexible tactile array sensor that provides pressure feedback. Section 5 describes the experimental setup and methods for terrain classification. Sections 6 and 7 gives the experimental results and the conclusion of this work, respectively.

2   Robot Leg with a Bio-inspired Compliant Foot for Walking on Cylindrical Terrains

We have developed a robot leg based on a hind leg of the dung beetle which has an interesting structure for both locomotion and curved object transportation. The leg has three active joints and three segments (coxa, femur, and tibia) between the joints. The segments are simplified and designed by following the proportion of the hind leg (i.e., coxa:femur:tibia is 1:1.2:1, see Fig. 1(a) and [1] for more details). The lengths of the coxa, femur, and tibia parts are 7 cm, 8.4 cm, and 7 cm, respectively. They are printed using 3D-printing. The CT- and FT-joint rotate around the z-axis while the TC-joint rotates around the x-axis. These rotations follow the joint rotations of the real dung beetle leg. The base of the TC-joint is attached to a linear slide allowing the leg freely move in a vertical direction during a stance phase. A flexible cable is used to hold the leg during a swing phase for ground clearance. We have simplified the robot foot by using a fin-ray inspired concept which compliancy mimics the segmented structure of the real tarsus of the beetle. It consists of five rays/blades embedded inside its triangular structure (Fig. 1(a)). It is printed using 3D-printing with a compliant material (i.e., rubber). Although this design does not fully capture the complete complex structure of the tarsus of the dung beetle, it, as an abstract version, shows flexibility and compliance to passively adapt its shape to follow the contour of a substrate as observed in the beetle. Besides the passive adaptation, the compliant foot also acts as a damping system to reduce contact force when the leg touches the ground, see more in [1].

3   Neural Locomotion Control

The concept of central pattern generators (CPGs) for locomotion has been studied and used in several robotic systems of particular walking robots. There is a wide variety of different CPG models available ranging from detailed biophysical models to pure mathematical oscillator models. Here, the model of a CPG for basic locomotion of robot leg is realized by using the discrete-time dynamics of a simple 2-neuron oscillator network (Fig. 1(b)). Due to its neurodynamics, it is able to autonomously generate various periodic and chaotic signals without sensory feedback; i.e., it can act as open-loop control. For our implementation here, the activity of each neuron develops according to $a_i(t+1) = \sum_{j=1}^{n} W_{ij}o_j(t)$; $i = 1, ..., n$ with an update frequency of 20 Hz, where n denotes the number of units. The neuron output o_i is given by a hyperbolic tangent (tanh) transfer function $o_i = \tanh(a_i) = \frac{2}{1+e^{-2a_i}} - 1$. W_i is the synaptic strength of the connection from neuron j to neuron i. The two neurons $H_{0,1}$ of the CPG are fully connected with the four synapses W_{00}, W_{01}, W_{10}, W_{11} and can form an oscillator if the weights are chosen according to an SO(2)-matrix:

$$\mathbf{W} = \begin{pmatrix} W_{00} & W_{01} \\ W_{10} & W_{11} \end{pmatrix} = \alpha \begin{pmatrix} cos(\varphi) & sin(\varphi) \\ -sin(\varphi) & cos(\varphi) \end{pmatrix}, \qquad (1)$$

with $-\pi < \varphi < \pi$ and $\alpha > 1$, the oscillator generates sine-shaped periodic outputs $o_{0,1}$ of the neurons $H_{0,1}$ (Fig. 1(c)) where φ defines a frequency of the output signals. In order to achieve stable locomotion (or stepping pattern), we

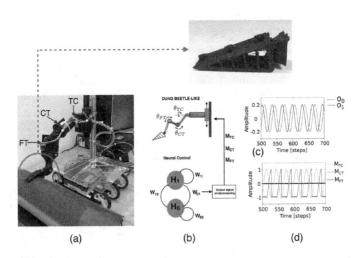

(a)   (b)   (d)

Fig. 1. (a) A robot leg with bio-inspired (fin ray) compliant foot. (b) CPG-based neural control for locomotion. It consists of two interconnected neurons $H_{0,1}$. (c) Outputs $o_{0,1}$ of the neurons $H_{0,1}$ of the CPG-based control. (d) Motor signals $M_{TC,CT,FT}$ obtained from a CPG output signal postprocessing unit. The post processing unit translates the outputs $o_{0,1}$ into the proper motor signals.

here set φ to 0.5 and α to 1.01 and use a CPG postprocessing unit to shape the CPG signals. The resulting signals $M_{TC,CT,FT}$ drive the motors of the leg (Fig. 1(d)). With this setup, the neural controller acts as an open-loop controller to control the leg. We use this control setup to generate a walking pattern of the leg on cylindrical terrains investigated here [1].

4   Flexible Tactile Array Sensor

The piezoresistive material has been chosen as the most suited to build a flexible tactile-array sensor. The tactile sensor is built as an array of 10×25 taxels (tactile cells) within $16\,\mathrm{mm} \times 40\,\mathrm{mm}$ responsive to pressure applied normally. A range of other sensor configurations were built previously and detailed information can be found in [2,15]. Each tactile cell exhibits decreased resistance for increased pressure with the resistance varying from 1 MΩ in the uncompressed state to under 1 kΩ in the stressed state. The pressure tactile sensor is composed of 3 layers overlaid, where the middle layer is a piezoresistive rubber, while the upper and lower layers being rows and columns of flex printed PCB wired perpendicularly. A multiplexing scheme based on the voltage divider principle implemented with dedicated electronics addresses all combinations of rows and columns and transforms them into a tactile image of 250 elements, each value being an 8-bit value. A second layer is added to the pressure sensor for sensing curvature based on the Spectra Symbol technology for measuring flexing of the material. Bending this sensor increases the resistance which is added to the tactile image of pressure information. The dedicated electronics module stream the information over wifi providing 30 frames per second of distributed pressure together with curvature information as shown in Fig. 2(a).

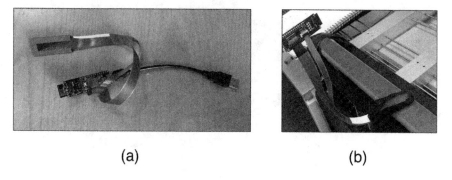

(a)   (b)

Fig. 2. (a) A flexible wireless tactile array sensor. (b) A compliant foot with a tactile array sensor covered by a rubber glove.

5   Experimental Setup and Methods for Terrain Classification

We implemented a compliant robot foot and a tactile array sensor together by fixing a tactile array sensor under a compliant foot. The robot foot was covered

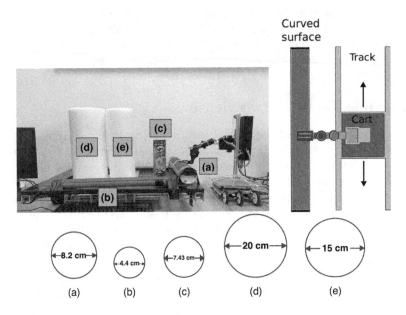

Fig. 3. Experimental set-up of the leg system for locomotion of the first experiment on cylindrical terrains which are (a) PVC A, (b) PVC B, (c) Hard Paper, (d) Sponge A, and (e) Sponge B.

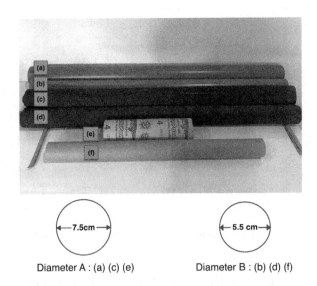

Fig. 4. Six different kinds of cylindrical terrains of the second experiment which are (a) PVC A, (b) PVC B, (c) Sponge A, (d) Sponge B, (e) Hard Paper A, (f) Hard Paper B.

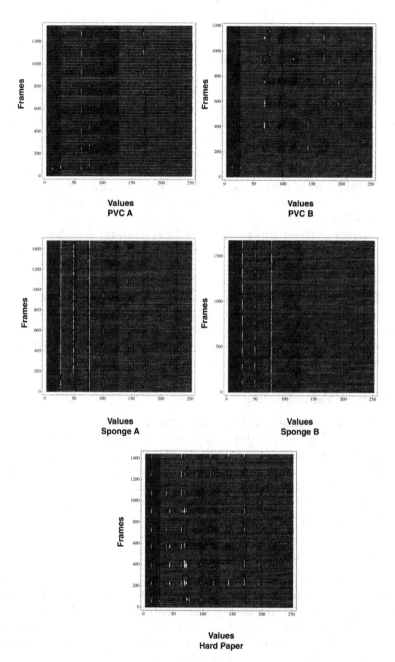

Fig. 5. Examples of heat maps of the tactile array sensor on each step while the compliant robot foot walks on each terrain. We encourage the reader to see the videos of the experiments at http://www.manoonpong.com/SAB2018/. (Color figure online)

by a rubber glove for having a friction while the robot is walking as shown in Fig. 2(b). The leg was attached to a moving cart which was constrained by two rails, to ensure that the leg moves along the terrains during locomotion [1]. We divided the experiment into two conditions. For the first condition, we provided the different diameters and different terrains. The setup of the leg with a cart to investigate locomotion efficiency on cylinder terrains with 3 different terrains and 5 different diameters which are a 8.2-cm diameter PVC cylinder (PVC A), a 4.4-cm diameter PVC cylinder (PVC B), a 7.4-cm diameter hard paper cylinder (Hard Paper), a 20-cm diameter sponge cylinder (Sponge A), and a 15-cm diameter sponge cylinder (Sponge B), was setup as shown in Fig. 3. The second condition, we provided the identical diameter with 3 different terrains which are a 7.5-cm diameter PVC cylinder (PVC A), a 5.5-cm diameter PVC cylinder (PVC B), a 7.5-cm diameter sponge cylinder (Sponge A), a 5.5-cm diameter sponge cylinder (Sponge B), a 7.5-cm diameter hard paper cylinder (Hard Paper A), and a 5.5-cm diameter hard paper (Hard Paper B) as shown in Fig. 4. In this experiment, the tests take 10 steps, 5 times on each terrain. For each terrain, 50 tests were done, averages on each terrain resulting in 50 trials in total. Figure 5 shows contact area on each step while the compliant robot foot walks on each terrain. We can determine the contact area during walking through pressure intensity on the heat maps. The green colour shows low pressure, the blue colour shows medium pressure, and the white colour shows high pressure on each area. The red colour means no foot contact.

5.1   Average Value Method

According to the experiment, the tests take 10 steps, 5 times on each terrain. We calculated the averages of each step, which means we got 50 values of the averages on each terrain. We also calculated the standard deviation for calculating the expanded uncertainty using metrology method. The 50 average values have been divided into half (25:25). The second half was tested using the data of the first half. We demonstrated the correction of classification by doing the confusion matrix (Tables 1 and 3).

5.2   Maximum Value Method

In this method, we calculated the maximum values on each step, which means we also got 50 values on each terrain. We have calculated the standard deviation for calculating the expanded uncertainty using metrology method as well. The 50 maximum values have been divided into half (25:25). The second half was tested using the data of the first half. We demonstrated the correction of classification by doing the confusion matrix (Tables 2 and 4).

6   Experimental Results

On the first experiment, the average method classification shows some mistakes as shown in Table 1. There are two terrains have a percentage of correction under

Table 1. Confusion matrix of the first experiment of average value method for cylindrical terrain classification. The vertical axis represents the truth and the horizontal represents the output of the classification in percentage.

AV Method						
Object	PVC A	PVC B	Hard Paper	Sponge A	Sponge B	Unknown
PVC A	88	12	0	0	0	0
PVC B	8	72	0	0	20	0
Hard Paper	8	0	0	4	56	32
Sponge A	0	0	0	96	0	4
Sponge B	0	8	4	36	20	32

Table 2. Confusion matrix of the first experiment of maximum value method for cylindrical terrain classification. The vertical axis represents the truth and the horizontal represents the output of the classification in percentage.

MV Method						
Object	PVC A	PVC B	Hard Paper	Sponge A	Sponge B	Unknown
PVC A	91.67	8.33	0	0	0	0
PVC B	0	64	36	0	0	0
Hard Paper	0	20	80	0	0	0
Sponge A	0	0	0	92	4	4
Sponge B	0	0	0	28	68	4

50, noticeably the Hard Paper has a percentage of correction equal to 0. Moreover, there are three terrains have 32 % of unknown values which are unidentified because of the overlap of the values. Therefore, the average value method cannot be used for cylindrical terrain classification. By contrast, the maximum value method shows that (Table 2) all of the materials are able to classify by using this method, especially, on large diameter materials (PVC A and Sponge A). They have a percentage of correction over 90. On the second experiment, we divided the terrains into two groups, which are Diameter A (7.5-cm) group and Diameter B (5.5-cm) group. The average method classification shows the huge mistakes almost every terrains accept Sponge B and every terrains have the unidentified value as shown in Table 3. Therefore, the average value method cannot be used for cylindrical terrain classification. By contrast, the maximum value method shows that (Table 4) all of the terrains are able to classify by using this method. The clear diagonal for maximum value method confirms the successful recognition. Therefore, this method is suited for the cylindrical terrain classification. According to the classification, we found that for the same terrains different diameters, a gap of the averages of maximum value between these materials will be small. On the other hand, if terrains are different whether diameters are same or different, a gap of the averages of maximum value between these terrains will be outstanding. We also can identify the difference of terrains by

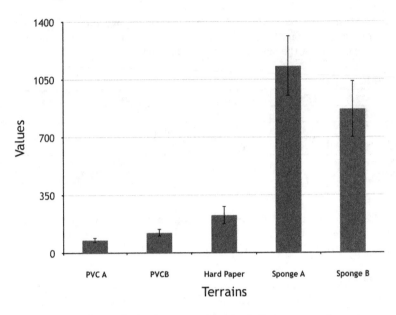

Fig. 6. Comparison chart of the first experiment of the average of maximum values while the compliant robot foot walks on each terrain.

Table 3. Confusion matrix of the the second experiment of average value method for cylindrical terrain classification. The vertical axis represents the truth and the horizontal represents the output of the classification in percentage.

AV Method				
Diameter A				
Object	PVC A	Hard Paper A	Sponge A	Unknown
PVC A	0	80	0	20
Hard Paper A	0	40	20	40
Sponge A	0	68	4	28

AV Method				
Diameter B				
Object	PVC B	Hard Paper B	Sponge B	Unknown
PVC B	12	0	60	28
Hard Paper B	40	0	56	4
Sponge B	20	0	72	8

hard terrains have an average maximum value less than soft terrains as shown in Figs. 6 and 7. For hard terrains, the bigger diameter has an average less than the smaller, but for soft terrains, the bigger diameter has an average more than the smaller. Moreover, we found that the more hardness of terrains, the less uncertainty. The reason is while a compliant robot foot is walking on hard terrains, there is no deformation of terrains but for soft terrains, there is a deformation on each step and the hardness on each point are different. The uncertainty of the average of maximum values was calculated by using metrology method for precise calculation.

Table 4. Confusion matrix of the second experiment of maximum value method for cylindrical terrain classification. The vertical axis represents the truth and the horizontal represents the output of the classification in percentage.

MV Method			
Diameter A			
Object	PVC A	Hard Paper A	Sponge A	Unknown
PVC A	81	19	0	0
Hard Paper A	10	65	25	0
Sponge A	0	25	75	0

MV Method			
Diameter B			
Object	PVC B	Hard Paper B	Sponge B	Unknown
PVC B	62	28	10	0
Hard Paper B	24	76	0	0
Sponge B	4	25	71	0

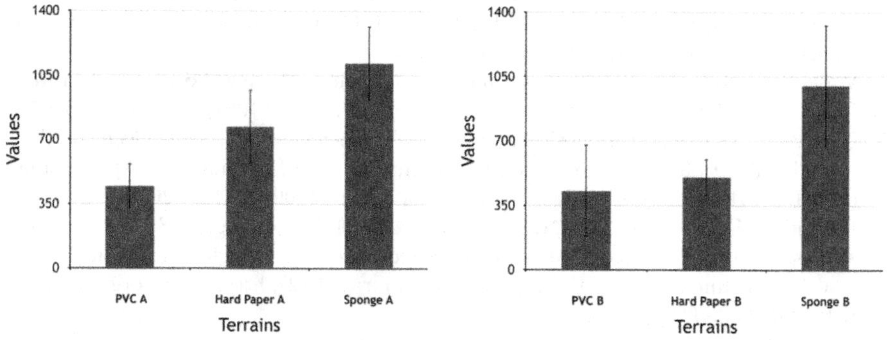

Fig. 7. Comparison chart of the second experiment of the average of maximum values while the compliant robot foot walks on each terrain.

7   Conclusion

In this study, we presented the combined usage of a compliant robot foot with a flexible tactile sensor array for cylindrical terrain classification. We performed two conditions in the experiment which are (1) different diameters and different terrains and (2) different terrains with the same diameter. Two different methods were used which are average and maximum value methods. We found that the suitable method for cylindrical terrain classification is maximum value method. We also can identify a different kind of materials or diameters by comparing the averages of the maximum value. In order to improve classification performance, two legs could be used as complementary to tactile sensing. Walking robots will be able to recognize terrains while they are walking, explore, and adjust their suitable walking pattern on different terrains in future.

Acknowledgements. This work was supported by the Capacity Building on Academic Competency of KU. Students from Kasetsart University, Thailand, Centre for BioRobotics (CBR) at University of Southern Denmark (SDU, Denmark), the Thousand Talents program of China, and the Human Frontier Science Program under grant agreement no. RGP0002/2017.

References

1. Di Canio, G., Stoyanov, S., Larsen, J.C., Hallam, J., Kovalev, A., Kleinteich, T., Gorb, S.N., Manoonpong, P.: A robot leg with compliant tarsus and its neural control for efficient and adaptive locomotion on complex terrains. Artif. Life Robot. **21**(3), 274–281 (2016)
2. Drimus, A., Kootstra, G., Bilberg, A., Kragic, D.: Design of a flexible tactile sensor for classification of rigid and deformable objects. Robotics and Autonomous Systems **62**, 3–15 (2014)
3. Mrva, J., Faigl, J.: Feature extraction for terrain classification with crawling robots. Robotics In: Yaghob, J. (ed.) ITAT 2015, 2010 Current and Future, pp. 179–185. Charles University in Prague, Prague (2015)
4. Walas, K.: Terrain classification and negotiation with a walking robot. Intell. Robot. Syst. **78**(3–4), 401–423 (2015). https://doi.org/10.1007/s10846-014-0067-0
5. Degrave, J., Van Cauwenbergh, R., wyffels, F., Waegeman, T., Schrauwen, B.: Terrain classification for a quadruped robot. In: Machine Learning and Applications (ICMLA). IEEE, Miami (2013). https://doi.org/10.1109/ICMLA.2013.39
6. Bermudez, F.L.C., Julian, C., Haldane, W., Abbeel, P., Fearing, R. S.: Performance analysis and terrain classification for a legged robot over rough terrain. In: Intelligent Robots and Systems, IEEE/RSJ, Vilamoura (2012). https://doi.org/10.1109/IROS.2012.6386243
7. Belter, D., Skrzypczyński, P.: Rough terrain mapping and classification for footholdselection in a walking robot. Field. Robot. **28**(4), 497–528 (2011). https://doi.org/10.1002/rob.20397
8. Kisung, K., Kwangjin K., Wansoo, K., Seungnam, Y., Changsoo H.: Performance Comparison between neural network and SVM for terrain classification of legged robot. In: SICE Annual Conference 2010, pp. 1343–1348. IEEE, Taipei (2010)
9. Poppinga, J., Birk, A., Pathak., K.: Hough based terrain classification for realtime detection of drivable ground. Field. Robot. **25**(1), 67–88 (2008). https://doi.org/10.1002/rob.20227
10. Wu, X.A., Huh, T.M., Mukherjee, R., Cutkosky, M.: Integrated ground reaction force sensing and terrain classification for small legged robots. IEEE Robot. Autom. Lett. **1**(2), 1125–1132 (2016)
11. Walas, K.: Tactile sensing for ground classification. J. Autom. Mobile Robot. Intell. Syst. **7**(2), 18–23 (2013)
12. Gong, D., He, R., Yu, J., Zuo., G.: A pneumatic tactile sensor for co-operative robots. Sensors **17**(11), 2592–2606 (2017)
13. Jamali, N., Sammut, C.: Material classification by tactile sensing using surface textures. In: Robotics and Automation, pp. 2336–2341. IEEE, Anchorage (2010). https://doi.org/10.1109/ROBOT.2010.5509675
14. Nakamoto, H., Kobayashi, F., Kojima, F.: Shape classification using tactile information in rotation manipulation by universal robot hand. In: Abdellatif , H. (Ed.) Robotics 2010 Current and Future Challenges. ISBN: 978-953-7619-78-7
15. Drimus, A., Mátéfi-Tempfli, S.: Tactile shoe inlays for high speed pressure monitoring. In: Liu, H., Kubota, N., Zhu, X., Dillmann, R., Zhou, D. (eds.) ICIRA 2015. LNCS (LNAI), vol. 9245, pp. 74–81. Springer, Cham (2015). https://doi.org/10.1007/978-3-319-22876-1_7

Action Selection and Navigation

An Artificial Circadian System for a Slow and Persistent Robot

Matthew J. O'Brien$^{(\boxtimes)}$ and Ronald C. Arkin

Mobile Robot Lab, Georgia Institute of Technology, Atlanta, GA 30308, USA
{mjobrien, arkin}@gatech.edu

Abstract. As robots become persistent agents in natural, dynamic environments, the ability to understand and predict how that environment changes becomes more valuable. Circadian rhythms inspired this work, demonstrating that many organisms benefit from maintaining simple models of their environments and how they change. In this work, we outline an architecture for an artificial circadian system (ACS) for a robotic agent. This entails two questions: how to model the environment, and how to adapt robot behavior based on those models. Modeling is handled by treating relevant environment states as time series, to build a model and forecast future values of that state. The forecasts are considered special percepts, a prediction of the future state rather than a measurement of the current state. An ethologically-based action-selection model incorporates this knowledge into the agent's decision making. The approach was tested on a simulated precision agricultural task - pest monitoring with a solar powered robot - where it improved performance and energy management.

1 Introduction

Moving from structured, static environments to natural, dynamic environments remains a challenge in robotics. While reactive approaches in behavior-based designs have shown effectiveness in dealing with the randomness present in the real world, most ignore the somewhat predictable dynamics that many environments exhibit. In this work we describe an architecture for modeling environmental dynamics, and adapting an agent's behavior to them. This approach is inspired by circadian rhythms found in a variety of natural organisms; from humans, to plants, to bacteria [15].

Circadian systems are chemical oscillators, synchronized to the solar cycle, that influence behavior and metabolic processes of an organism. Rather than simply reacting to the environmental changes, circadian systems allow an organism to take action before its senses (perceptual state) or needs (internal state) could drive the behavior. "Circadian rhythmicity of behavior represents an animal's information, or one is tempted to say, knowledge about a particular feature of its environment... and what to do about it." [14]

This research is supported by The Office of Naval Research Grant #N00014-15-1-2115.

© Springer Nature Switzerland AG 2018
P. Manoonpong et al. (Eds.): SAB 2018, LNAI 10994, pp. 149–161, 2018.
https://doi.org/10.1007/978-3-319-97628-0_13

The need for such an architecture in a robotic system is growing. Recently, the non-industrial robotic market overtook the industrial robotic market for the first time [18]. With the influx of robots into our lives, interest and research into persistent autonomy is increasing. Particular focus has been placed on mapping dynamic environments. While some model those dynamics [1], many focus on ways to filter changes to update the map [7]. More relevant is work on modeling environment dynamics that is applied to robot behavior. The most similar work modeled binary environment states as a periodic probability using Fourier series, and applied it to several problems including path planning, localization, and exploration [11].

This work is particularly targeted for slow, persistent robotic agents. Persistent agents will experience environmental cycles that a robot that executes for twenty minutes, or even two hours, will not. Slow robots are inherently less able to react quickly to a changing environment (but may have several advantages, especially for persistent energy-constrained tasks [3]). Thus, the ability to predict and proactively act could be very beneficial.

In this research, we develop an artificial circadian system (ACS) to learn and exploit the patterns that often exist in the dynamics of natural environments. In previous work, we demonstrated the application of time series modeling of an environment for a robotic agent, and tested it on a simulated robot interacting with pedestrian traffic using very simple behavioral rules [13]. Here, we investigate a principled way to incorporate the generated predictions into action-selection.

2 The Architecture

The ACS is built on a behavior-based architecture. A special set of perceptual schemas (or perceptual algorithms that process sensor data for specific behaviors [2]) model the environmental dynamics as time series and forecast future values of the state. Behaviors are supplemented with activation functions based on both the current sensed state and the future predicted state. Action-selection is done by selecting the behavior with the highest activation. Figure 1 shows the top-level architecture, and the rest of this section details the components.

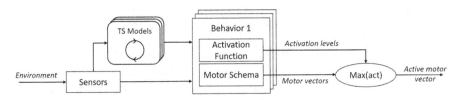

Fig. 1. The top-level diagram of the ACS architecture.

2.1 Modeling Environmental Dynamics as Time Series

We view the environment state as a time series, and apply methods from the time series literature to model and forecast future values. Time series models predict, or "forecast", future values of a variable directly from its past values. It allows for modeling of both cyclic and non-cyclic effects, and can forecast both the value of the future state as well as generate prediction intervals. A useful quality is that the approach is data driven. The underlying cause of the dynamics may be too complicated to model, or at least impractical to do so on a robot (consider the actions of many individuals to create traffic). Time series modeling offers versatility and simplicity, side-stepping the need for expert knowledge about a domain.

A fundamental approach to modeling time series is the classical decomposition [8], of a time series into its major components:

$$Y_t = T_t + S_t + E_t \tag{1}$$

Where Y_t is the original time series. T_t is the trend component, a slowly changing average level; S_t is the seasonal component, a repeating pattern with known period; and E_t is the residual, or error, left over after the trend and seasonal components have been removed from the time series. This decomposition provides a structured approach for modeling and allows for focus on components of interest. For example, seasonal effects are sometimes removed from data in a process called deseasonalization. In our work the seasonal effects are of key importance, as they represent the cyclic (or circadian, if the period is roughly 24 h) dynamics in the environment.

The literature on time series analysis and forecasting provides a broad set of tools [9]. At this time, however, there is no catch-all method or system. Historical data is required to manually select an appropriate model, one that can capture the dynamics of the environment. While model selection is must be done offline, model fitting (i.e. parameter estimation) and forecasting can be done online, autonomously by the agent.

Time series modeling has traditionally been applied for offline analysis of discrete data. To leverage these techniques on a real-time system, three data structures are used when updating the forecast. The past state values, or the measured time series, is stored as the vector S_p. These are fed into the time series model, and used to forecast the future values of the state, S_F^t, where t is the time this forecast vector was generated. Lastly, there is a set of current sensor measurements called S_M. At the end of any forecast interval, S_M is used to generate a final state value (S_p) for that interval. This could be an average of multiple measurements, a count of recorded events, or even a null value if no measurements of the relevant state variable were made (Fig. 2).

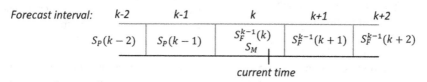

Forecast interval:	k-2	k-1	k	k+1	k+2
	$S_P(k-2)$	$S_P(k-1)$	$S_F^{k-1}(k)$ S_M	$S_F^{k-1}(k+1)$	$S_F^{k-1}(k+2)$

current time

Fig. 2. S_P returns measures of the state in the *past*, while S_F returns forecasts of the state in the future. S_M is the current measured state.

Entrainment of this system to the environment happens at two levels. Every forecast interval the agent appends a new value to S_P, and generates a new forecast vector S_F. This means the forecast for some specific point in time in the future is repeatedly updated (usually becoming more accurate) as new measurements of the environment are taken. Secondarily the new recent history, S_P, can be used to re-estimate the model parameters. Adapting the time series model if the dynamics of the environment have changed.

2.2   Action-Selection via Activation Levels

The action-selection method used in this work is part of a family of approaches that assigns an activation level to each top-level behavior an agent can execute. Arbitration via action-selection is straightforward: the behavior with the highest activation level is executed. This leaves the question of how to calculate that activation level. Many approaches for this problem have been developed [5, 16, etc.]. The approach in this paper is based on an ethologically-inspired architecture, described in [4].

Every behavior includes several additional components to facilitate the action-selection process (Fig. 3). The first is a motivation function (MOT) which represents a behavior's internal motivation to activate. This is based on endogenous (internal) variables representing the state of the agent (such as its power level). The second component is the releasing mechanism (RM). It differs from the motivation function in that it focuses on external, or exogeneous, variables. The RM is a special function that specifies both the conditions required for a behavior to activate, and how well they are satisfied. Algorithmically, if its output is zero it means necessary conditions to execute a behavior are not currently satisfied, and the behavior cannot execute regardless of how high its total activation may be. In this research, we introduce a new component, the circadian

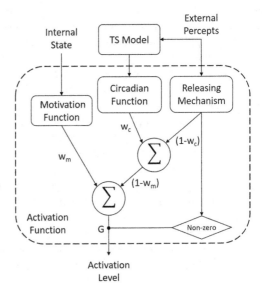

Fig. 3. The activation function of a behavior is a weighted sum of the three components, plus a conditional check on the releasing mechanism

rhythm function (CIR). This component represents the influence of an associated forecast on a behaviors activation level. Like the releasing mechanism, the circadian rhythm function focuses on exogenous variables, but considers the impact of their future predicted values, rather than their current measured value.

Each behavior has an activation function that is the weighted sum of the motivation function, releasing mechanism, and circadian rhythm function. The activation function produces a final activation level used to select which behavior to execute. The activation function of behavior N is:

$$
act_N = \begin{cases} 0 & \text{if RM conditions not satisfied} \\ G * (w_m * MOT_N + (1 - w_m)(w_C * CIR_N + (1 - w_C) * RM_N)) & \text{else} \end{cases} \tag{2}
$$

Three parameters are introduced to conveniently adjust robot behavior, based on the approach taken in [6]. The behavior gain, G, adjusts the overall activation magnitude of a behavior. The weight $\{w_M \in R \vee 0 \leq w_M \leq 1\}$ adjusts the relative influence of the releasing mechanism and the motivation function. This allows for the robot to become more reactive to its environment (higher weighting of releasing mechanism and circadian rhythm function) or more focused on its internal needs and goals (higher weights of motivation function). A final weight is introduced, $\{w_C \in R \vee 0 \leq w_C \leq 1\}$. This weight balances the impact of the CIR function against the releasing mechanism, or more generally sets how much the robot focuses on the current measured state compared to the predicted future state.

The circadian function allows for more than acting directly on a forecasted state. For instance, choosing an ideal time to execute a behavior over some time window can be done by comparing the current state against the forecasted state throughout that window. An otherwise hidden state can be estimated using forecasts generated in the past of the now current state. In the following experiment, both of these approaches are leveraged.

3   Experimental Validation

To validate this architecture, we set up a simulated experiment to test how incorporating the circadian rhythm function impacts an agent's behavior. The scenario involves a small, slow agriculture robot; one that persists within the agricultural environment as a beneficial component of the ecosystem, monitoring and tending to the plants.

The agent's purpose is to monitor the population of a pest over a growing season, and identify when the population reaches a critical threshold requiring intervention by a farmer. It must do so while spending a minimal amount of time and energy, allowing the agent to work on other tasks, and reducing wear on physical components. The dynamics modeled by the time series will be the pest population and solar energy. This will inform when the agent must spend more time and energy on monitoring the pest population, and when it should charge.

3.1   Environment Details and Modeling

The simulation is implemented in Gazebo as shown in Fig. 4. The pest population dynamics follow a model based on aphids. Many aphid species exhibit exponential population growth after an initial infestation due to their ability to perform rapid asexual reproduction during some periods of their life cycle. This growth is simulated by inserting pests according to a simple aphid population model [10], plus a random noise factor. Pest are distributed uniformly over the work space. Figure 5 shows several simulated trajectories of the aphid population. The robot is solar charged, and must spend a significant portion of its time

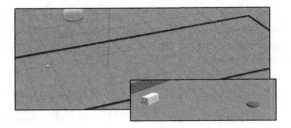

Fig. 4. The Gazebo simulation environment. The bottom right image highlights the robot (white box) and pest (red disk) (Color figure online)

charging. It expends energy whenever moving, proportional to its speed. The robot receives "direct" solar light when against the top wall closest to the "sun" (yellow sphere), and reduced solar light as it moves away. Available solar power over time followed a bell shape plus noise between dawn and dusk, and was zero at night time.

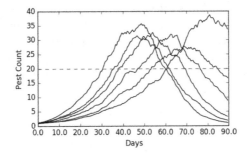

Fig. 5. Example population trajectories used for simulation. The critical threshold for pest population, where intervention is needed, is shown as a dashed red line. Different trials reached this threshold over 30 days apart. (Color figure online)

The pest dynamics were stored as a time series of measured pest levels, defined as the number of pests found per area searched. A 12-h time interval was used. As the agent often did not monitor in a 12-h period, values between measured intervals were interpolated, while values after any measurement were forecasted. The pest dynamics were modeled using the STL decomposition (Seasonal and Trend decomposition using Loess) in the R forecasting library [9] which breaks the series into a trend, seasonal, and stationary components. For details on these forecasting techniques, see [8]. Nine previous seasons, each 90 days long, were simulated to provide data to fit the model parameters.

Solar energy is this simulation was simple, a set cyclic function over 24 h, plus noise. Thus, it was forecasted as the average or expected solar level per time step.

3.2   Robot Behaviors

While the top-level action-selection is handled by the architecture described in Sect. 2, each behavior is built as an assemblage of simpler behaviors. The behaviors are based on work exploring the design of slowbots (i.e. slow robots [17]). The monitor behavior is responsible for pest monitoring. It drives the robot around its work space, exploring and searching for pests. This is implemented as a vector summation of three behaviors: a wander behavior to drive the robot around its work space, an attachment behavior [12] to keep the robot in its workspace, and an obstacle avoidance behavior.

For the activation function, the monitor behavior's MOT (Eq. 3 below) drives the agent to monitor the environment at periodic intervals. When not monitoring, its activation level increases proportional to the time since it last monitored. When monitoring, the activation level decreases proportional to the time is has been recently monitoring (we used a 12-h window). The RM (Eq. 4) increases activation of the monitor behavior based on how much the forecasted pest level, and the last measured pest level, disagree. This causes the robot to monitor longer and more frequently when the current pest population deviates from expected values. The RM has a small constant factor to always release the monitor behavior, as the environment is always available to monitor. The CIR (Eq. 5) increases activation proportional to the expected level of pests. This is the key component that prioritizes the robot's actions when there are more pests, or a higher expected number of pests.

In the below equations, T_S is the time (minutes) since the monitoring behavior last executed. T_M is the time (minutes) spent monitoring in the last 12 h. P_M is the last measured pest level, and $P_F(t)$ is the forecasted pest level for time t. All K_n values are constant parameters.

$$MOT(T_M, T_S) = \begin{cases} K_1 * T_S & \textit{if monitor inactive} \\ 100 - K_2 * T_M & \textit{if monitor active} \end{cases} \qquad (3)$$

$$\textit{constrained such that} \qquad 0 \leq MOT(T_M, T_S) \leq 100$$

$$RM(P_M, P_F(t)) = K_3 * |P_M - P_F(t)| + 0.1 \qquad (4)$$

$$CIR(P_F(t)) = K_4 * P_F(t) \qquad (5)$$

The energy collection behavior moves the robot to the region of its work space where direct sunlight is available. Once there the agent waits, charging, until another behavior takes over executing. This is implemented through a simple FSA (see Fig. 6).

The energy collection behavior's motivation function (Eq. 6) is a function of the internal battery level, and it increases exponentially as the robot's battery level decreases. This was arbitrarily chosen, but ensures the behavior becomes dominant when power is low. The RM (Eq. 7) increases the activation level proportional to the amount of solar energy currently available to the robot, averaged over the recent history. The behavior is released even when there is no sunlight so that the agent can move and wait for energy, rather than monitor when its battery might be too low. Finally, the CIR (Eq. 8) looks at the average solar energy for a set time in the future

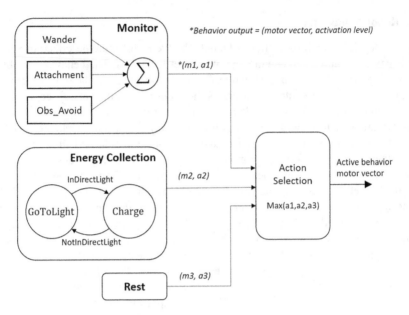

Fig. 6. Top-level behavioral architecture

(in this experiment, 90 min), and compares that to the current solar energy. When the current solar energy is higher, the activation level of energy collection is increased, making the robot more likely to charge before the opportunity is lost.

In the following equations, B is the battery level of robot (as a percentage), E_M is the current measured solar energy, and $E_F(t)$ is the forecasted solar energy at some point in time, t.

$$MOT(\text{B}) = K_5 * 10^{1-B} - K_6 \tag{6}$$

$$RM(E_M) = K_7 * E_M + 0.1 \tag{7}$$

$$CIR(E_M, E_F(t)) = \begin{cases} K_8 * \left(E_F(t) - E_{avg}\right) & \text{if } E_F(t) - E_{avg} > 0 \\ 0 & \text{else} \end{cases} \tag{8}$$

$$E_{avg} = \frac{1}{N} \sum_{i=t}^{i=t+N} E_F(i) \tag{9}$$

The final behavior, rest, simply causes the robot to stop moving. This behavior exists as an alternative to always attempting to collect energy or monitor, even when the activation of both behaviors is low. Thus, if there is nothing important for the agent to do, the agent will do nothing. The agent will still charge if its positioned in direct sunlight while resting. The rest behavior maintains a constant activation level at all times, achieved by a constant value for the MOT, RM, and CIR functions. There is potential for a more sophisticated activation function, potentially taking into consideration over heating or bad weather, but this was not the focus of this work. Figure 6 shows the top-level diagram of the robot's architecture for this experiment.

3.3 Experiment Procedure

The robot's task was to detect when the pest population crossed a critical threshold. This goal was achieved by executing the monitoring behavior and measuring the pest population after said threshold is reached. The experiment varied the weight of the CIR function for both behaviors (monitor and charge) across a range of values. For performance with respect to the goal, the *time-to-detection* measure is defined as the time between when the critical threshold was reached, and when the robot detected it. With respect to energy, two others are defined. *Energy-spent* is the energy used by the robot in one trial. *Charge-rate* is the rate of energy collected, per minute, during the periods the robot was charging. These measure the efficiency of both working and charging in terms of energy.

The experiment was executed 20 times for each condition. Trials had different initial conditions, pest locations, and pest population growth included some randomness. The hypotheses for the experiment was that the inclusion of the CIR component in the activation function of the monitor behavior will cause the robot to detect that the pest population reached the threshold sooner (lower *time-to-detection*) while spending less time/energy in the monitoring behavior (lower *energy-spent*). For the charge behavior, the CIR component will improve the timing of when the robot charges (higher *charge-rate*).

4 Results and Discussion

4.1 The Monitor Behavior

When testing the monitor behavior, the charge behavior weights were set (by experimentation) to: $W_M = 0.5$, $W_C = 0.3$. The monitor activation weights were $W_C = 0.9$, and W_M varied from 1.0 (only use MOT) to 0.4 (heavily use CIR and RM components). Figures 7 and 8 show the performance for *time-to-detection* and *energy-spent*. There is a clear trend that higher weighting of the CIR component reduced the average *time-to-detection*, reducing it by **85.9%** between 0% and 54% CIR weights (2267 min to 319.9). The time-to-detection also has significant variance, as the robot could randomly monitor right before or after the pest threshold is reached. The standard deviation was reduced by **74%** with higher weighting of the CIR: from 1352 min to 351.

Figure 9 illustrates the difference in behavior. Without the CIR component, the agent uses no knowledge about the pest population, and monitors at regular intervals. In this case, the ACS predicts an otherwise hidden variable: the level of pests when the robot is not monitoring. This allows the agent to use less energy to monitor when the pest threat is low, but more energy to monitor near the critical time when pests are becoming a problem. The average energy used per day was minimized at 36% CIR weight, where it was **22.2%** lower than with 0% CIR weight. With higher CIR weight,

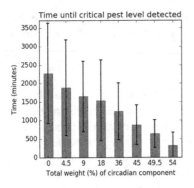

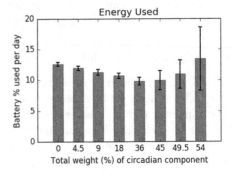

Fig. 7. Average *time-to-detection* over 20 trials when monitor behavior weights varied. One standard deviation marked.

Fig. 8. Average *energy-spent* per day over 20 trials when monitor behavior weights varied. One standard deviation marked.

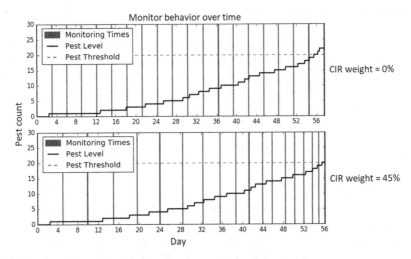

Fig. 9. Two executions by the agent on the same trial with different activation function weighting. This highlights the difference between when the agent monitors over the course of the experiment if the CIR component is included in the activation function.

the overall energy usage begins to increase, up to **6.71%** above the value at 0% CIR weight. Once the CIR weight reaches high enough, the agent may begin to monitor continuously (if it has energy to do so). This depends on the forecasted pest level, it did not happen in every trial, which creates the rapidly growing variance in energy usage. This effect happened when the CIR component alone can surpass the rest behavior in terms of activation. This significantly reduces the average *time-to-detection*, but expends a large amount of energy to do so.

4.2   The Charge Behavior

When testing the charge behavior, the monitor behavior was held constant with weights $W_M = 0.5$ and $W_C = 0.9$. The charge behaviors weights were adjusted with a slightly different scheme. The releasing mechanism was an important component to avoid behavioral dithering (or rapid switching). It was left at a total 30% of the activation weight, while the total weight of the CIR component was varied from 0% to 40%. W_M and W_C were both varied to achieve these ratios of the MOT and CIR components on final activation level.

The *charge-rate*, as shown in Fig. 10, has a trend of increasing with greater weight of the CIR component: from an average 0.1778 charge-% per min to 0.4187, an increase of **135%**. The charge behavior's circadian rhythm function particularly prioritizes charging near peak sunlight, before the ideal opportunity to charge is lost. Thus, the agent is reacting not just on its current power and the current solar energy, but also on how the solar energy is changing. This has a clear impact on performance. The *time-to-detection* was also investigated for the charge behavior testing (Fig. 11) to see if there might be effects on performance due to the changes to charge timings. However, no trend seems to have emerged.

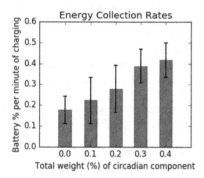
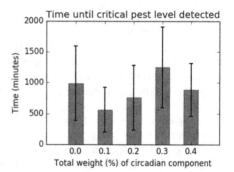

Fig. 10. Average *charge-rate* over 20 trials when charge behavior weighs varied. One standard deviation marked.

Fig. 11. Average *time-to-detection* over 20 trials when charge behavior weights varied. One standard deviation marked.

5   Conclusions

This work has detailed an architecture to help a robotic agent respond to long-term dynamics - the Artificial Circadian System. Inspired by the circadian systems in nature, time series models forecast the changing environment. These forecasts are incorporated into the action-selection process to allow a robot to act based on both the current measured and future predicted values of the state. The time series models can also be applied to estimate an otherwise hidden state, or detect deviations in the environment by comparing an expected and measured state.

An agricultural domain was chosen as a test bed, and a pest-monitoring task with a solar-powered robot was simulated. With the ACS, the agent was able to model the pest level over time, even when not actively monitoring it. Results showed this allowed for quicker identification of when the pest level reached a critical threshold requiring action, while using less energy to do so. The agent also weighted the future predicted availability of solar energy into its decisions on when to charge, improving the timing of transitions to a charging behavior. The agent's behavior adapted to the changing environment: monitoring more when the pest level was higher, and prioritizing charging when solar energy was at its peak.

Future work is planned in several directions. The next step is to leave simulations and begin testing with physical robots interacting with real environments. Robustness should be developed to handle cases when the forecasting system fails, due to modeling error or random events, allowing the agent to detect and respond to these situations. Finally, reinforcement learning offers interesting possibilities for the agent to autonomously adapt even when the environmental dynamics themselves change, but also brings significant challenges. A single state becomes a multi-dimensional trajectory when forecasted, significantly increasing the complexity of the state space.

References

1. Ambrus, R., Ekekrantz, J., Folkesson, J., Jensfelt, P.: Unsupervised learning of spatial-temporal models of objects in a long-term autonomy scenario. In: Intelligent Robots and Systems (IROS 2015), pp. 5678–5685. Curran Associates, Inc (2016)
2. Arkin, R.C.: Behavior-based robotics. MIT press, Cambridge (1998)
3. Arkin, R.C., Egerstedt, M.: Temporal heterogeneity and the value of slowness in robotic systems. In: IEEE International Conference on Robotics and Biomimetics (ROBIO 2015), pp. 1000–1005. Curran Associates, Inc (2016)
4. Arkin, R.C., Fujita, M., Takagi, T., Hasegawa, R.: An ethological and emotional basis for human–robot interaction. Robot. Auton. Syst. **42**(3), 191–201 (2003). https://doi.org/10.1016/S0921-8890(02)00375-5
5. Blumberg, B.: Action-selection in hamsterdam: lessons from ethology. In: Proceedings of the Third International Conference on Simulation of Adaptive Behavior: From Animals to Animats, pp. 108–117. MIT Press (1994)
6. Chernova, S., Arkin, R.C.: From deliberative to routine behaviors: a cognitively inspired action-selection mechanism for routine behavior capture. Adapt. Behav. **15**(2), 199–216 (2007). https://doi.org/10.1177/1059712306076255
7. Dayoub, F., Duckett, T.: An adaptive appearance-based map for long-term topological localization of mobile robots. In: Intelligent Robots and Systems (IROS 2008), pp. 3364–3369. Curran Associates, Inc (2009)
8. Hyndman, R., Koehler, A.B., Ord, J.K., Snyder, R.D.: Forecasting with Exponential Smoothing: The State Space Approach. Springer Science & Business Media, Heidelberg (2008)
9. Hyndman, R.J.: Forecasting Functions for Time Series and Linear Models. R package (2017). http://pkg.robjhyndman.com/forecast/. Accessed Jan 2018
10. Kindlmann, P., Arditi, R., Dixon, A.F.G.: A simple aphid population model. In: Aphids in a New Millennium, pp. 325–330. INRA, Paris (2004)
11. Krajník, T., Fentanes, J.P., Santos, J.M., Duckett, T.: Fremen: frequency map enhancement for long-term mobile robot autonomy in changing environments. IEEE Trans. Rob. **33**(4), 964–977 (2017). https://doi.org/10.1109/TRO.2017.2665664

12. Likhachev, M., Arkin, R.C.: Robotic comfort zones. In: Sensor Fusion and Decentralized Control in Robotic Systems III, International Society for Optics and Photonics, vol. 4196, pp. 27–42 (2000). https://doi.org/10.1117/12.403722
13. O'Brien, M.J., Arkin, R.C.: Modeling temporally dynamic environments for persistent autonomous agents. In: Proceedings of the Thirtieth International Florida Artificial Intelligence Research Society Conference, pp. 442–448. The AAAI Press, Palo Alto (2017)
14. Oatley, K.: Circadian rhythms and representations of the environment in motivational systems. In: McFarland, D.J. (ed.) Motivational Control Systems Analysis, pp. 427–459. Academic Press, Oxford (1974)
15. Paranjpe, D.A., Sharma, V.: Evolution of temporal order in living organisms. J. Circadian Rhythms. 3(7) (2005). https://doi.org/10.1186/1740-3391-3-7
16. Richter, M., Sandamirskaya, Y., Schöner, G.: A robotic architecture for action selection and behavioral organization inspired by human cognition. In: Intelligent Robots and Systems (IROS 2012), pp. 2457–2464. Curran Associates, Inc (2012)
17. Velayudhan, L., Arkin, R.C.: Sloth and slow loris inspired behavioral controller for a robotic agent. In: IEEE-ROBIO, International Conference on Robotics and Biomimetics, Macau (2017)
18. The Robotics Industry Will Reach $237 Billion in Revenue Worldwide by 2022. Tratica, 5 July 2017. https://www.tractica.com/newsroom/press-releases/the-robotics-industry-will-reach-237-billion-in-revenue-worldwide-by-2022/. Accessed Jan 2018

A Hybrid Visual-Model Based Robot Control Strategy for Micro Ground Robots

Cheng Hu$^{(\boxtimes)}$ ⓘ, Qinbing Fu ⓘ, Tian Liu ⓘ, and Shigang Yue ⓘ

University of Lincoln, Lincoln, UK
{chu,qifu,tliu,syue}@lincoln.ac.uk

Abstract. This paper proposed a hybrid vision-based robot control strategy for micro ground robots by mediating two vision models from mixed categories: a bio-inspired collision avoidance model and a segmentation based target following model. The implemented model coordination strategy is described as a probabilistic model using finite state machine (FSM) that allows the robot to switch behaviours adapting to the acquired visual information. Experiments demonstrated the stability and convergence of the embedded hybrid system by real robots, including the studying of collective behaviour by a swarm of such robots with environment mediation. This research enables micro robots to run visual models with more complexity. Moreover, it showed the possibility to realize aggregation behaviour on micro robots by utilizing vision as the only sensing modality from non-omnidirectional cameras.

Keywords: Micro robot · Visual model · Bio-inspired
Collision avoidance · Image processing

1 Introduction

Bio-plausible visual models inspired from insects are getting importance in robotics for various of visual-motor tasks, such as the trajectory stabilization and navigation task inspired by Elementary Motion Detector (EMD) [1], collision avoidance (CA) inspired by Lobular Giant Movement Detector (LGMD) [2,3], homing task inspired by the mushroombody of ants [4] and also navigation in spired by Small Target Motion Detectors (STMD) [5], by taking advantage of their reliability in coping with rapid changing scenarios with only minimum amount of neurons occupied. Comparing to conventional reactive navigation algorithms based on computer vision techniques such as standard optic flow [6], these bio-inspired models are usually free from massive calculation such as object recognition or distance estimation. Benefiting from the simplicity and efficiency, they are especially feasible for micro robots that often have only limited computational power on-board [7,8].

Take collision avoidance as one example, the neural structure LGMD which located in the locust's optic lobe is believed to be responsible for detecting looming objects and triggering escaping maneuver [9]. For decades, it has

ⓒ Springer Nature Switzerland AG 2018
P. Manoonpong et al. (Eds.): SAB 2018, LNAI 10994, pp. 162–174, 2018.
https://doi.org/10.1007/978-3-319-97628-0_14

inspired researchers to implement the neural network into computational models for autonomous vehicles. The computational model of LGMD such as the Embedded-LGMD (ELGMD) [8] has been proved to be reliable in dealing with collision situations in dynamic scenarios. By producing neural spikes only when the imminent collision is rapid and close enough, the LGMD is well-recognized by its high non-linearity [10] and directional selectivity, which is different from other bio-inspired visual models [1,11].

As a result, it will be interesting to investigate the possibility and performance of utilizing both LGMD-based collision avoidance model and visual navigation models to perform more complex tasks. Since they usually hold contradictory purposes, when both collision avoidance and visual navigation models are utilized, their feature have to be carefully mediated to avoid conflicting motion commands. Among various of approaches at higher level to integrate these visual models, such as feedback linearisation [12], fuzzy logic [13] and finite state machine (FSM) [14], the FSM is one of the most favoured approach on micro robot platforms due to its predictability responding to inputs without altering the performance of sub-models.

This paper proposed an approach of realizing a hybrid visual-motor control strategy for micro robots, allowing them to accomplish multiple tasks with certain goal. The proposed strategy which can be described as probabilistic model using FSM is composed of visual models from two different categories: the ELGMD model and a visual navigation approach based on image segmentation to imitate the target following behaviour. The stability and convergence of this robot control strategy is demonstrated by a series of experiments, including aggregation behaviour from a swarm of low-cost micro ground robots named *Colias IV*. To the best of the author's knowledge, it is the first attempt of utilizing a common camera as the only sensor, and processing image information to implement aggregation behaviour without communication on real micro robots.

The rest of this paper are organized as follows: Sect. 2 describes the proposed algorithms implemented in individual robots; Sect. 3 illustrates a series of experiments to test the performance of proposed robot system and the swarm scenario; we conclude this research in Sect. 4.

2   Models and Methods

2.1   The Bio-Inspired Collision Detection Model

Avoiding collisions is always crucial for autonomous robots. In this study, the ELGMD is deployed to serve CA purpose. As demonstrated in previous studies [8,15], even this feature alone, an individual robot is able to establish autonomous motion inside a constrained arena.

The ELGMD is a layered neural model formed by five layers with lateral inhibition mechanism and two single cells. The computational model contains only low-level image processing such as excitation transferring and neighbouring operations. The ELGMD model predicts an imminent collision by abstracting the visual motion information, i.e. the fast expanding profile of approaching

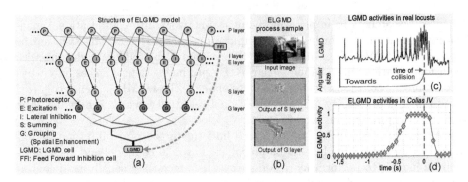

Fig. 1. The illustration of the implemented computation model ELGMD for CA. (a) The ELGMD model's architecture. The input of the *P layer* are the luminance change. Inhibitions (The *I layer* and the FFI cell) are indicated with dotted lines and have one frame delay. Excitations are indicated with solid lines which have no delay. (b) Example of ELGMD process. The output of *S layer* filters out the stationary background, and the output of *G layer* enhances the edges of moving foreground; (c) Intracellular recordings from the LGMD of a real locust viewing an approaching object (figure adapted from [9] with permission). (d) Output of ELGMD tested with a similar configuration comparing to (c), showing the increasing LGMD activity before collision.

objects into the activity level of LGMD cell, which is further transformed into neural spikes according to a fixed threshold. Confirmation of impending collision is generated by successive LGMD spikes. In some circumstances such as self turning, false spikes caused by sudden motion in vision could be suppressed by the feed-forward inhibition cell (FFI), which is also an activity from visual motion. The explanation of ELGMD and its performance are illustrated in Fig. 1. In the *Colias IV* robot, the process of ELGMD takes the full size (99×72) of captured image into calculation. For the robot's motion settings, since the speed characteristics is not the primary goal to study in this paper, the motion speed is not set to the maximum, but to the power-efficient range.

Table 1. Robot control strategy definition by ELGMD model

Neuron status		Decision	Action
C_f^{LGMD}	C_f^{FFI}		
0	0	No collision	Wander ($60\,mm/s$)
1	0	Collision	Turn ($180 \pm 60°$)
X(any)	1	Sudden change	Stop for a while ($0.3\ s$)

In consistency with ELGMD models utilized in autonomous CA experiments, the parameters and controlling logic are kept as previously chosen [8]. The robot's motion is controlled according to the status of ELGMD model, determined by both of the LGMD and FFI cells' activity, which are listed in Table 1. When no collision is detected, the robot is allowed to wander forward without constraints. To keep the system as simple as possible, the collision avoiding behaviour is defined by a in-position turning to another direction, which is almost an U-turn with randomly generated margin ($180 \pm 60°$) to prevent deadlocking in case trapped in the corner or several obstacles. When a large object is translating in front of the robot that producing spikes to the FFI cell, the robot will stop for a while until safe.

2.2 Vision Cue-Based Target Following Model

Another elementary part of this hybrid vision model is to track and follow an object, either a partner robot, or an anchor position, so that multiple robots could gather together and form an accumulating group. As one example of simple conventional visual model, it contains two parts: a light-weight visual target detection algorithm to identify the object, and a robot navigation method to follow the object continuously. Example of this process is shown in Fig. 2.

Visual target detection: The visual target detection algorithm is mainly based on the target's colour property. In this colour segmentation approach, we have chosen red as the only interested colour since it has the highest saturation response, which could be observed by the robot from far away.

Respecting to the robot platform's computational power, the algorithm is kept as simple as possible. The input image I used for this model has been down-sampled to 18×15 pixels to balance the image quality and the CPU&RAM occupation. The colour space is transformed into hue-saturation-value (HSV) [16] from it's original YUV format, thus majority of required chromatic cue is represented by the hue value. In HSV colour space, each pixel $I_{(i,j)}$ in the image I is represented by the values in three channels:

$$I(i,j) = [I_h(i,j),\ I_s(i,j),\ I_v(i,j)]'$$ (1)

At first, the down-sampling image is passes through a hue-saturation mask to match certain conditions. To prevent the hue adrift problem and to increase the robustness against different illumination conditions, suspicious pixels that belongs to the target are checked by the hue range and minimum saturation:

$$M_{red}(i,j) = \begin{cases} true & \text{if } (-10° \leq I_{h(i,j)} \leq -10°) \bigcap (I_{s(i,j)} \geq 0.3) \\ false & \text{otherwise} \end{cases}$$ (2)

where $M_{red}(i,j)$ is the suspicious region. The fixed thresholds are chosen empirically. Followed by this, all suspicious regions are joined and analysed by 4-connection to find the correct target region. Only the largest region $\mathbf{R}_r$ is treated the correct target area, excluding the others. Once a red object is identified by above procedures, two metrics are measured for the motion generation

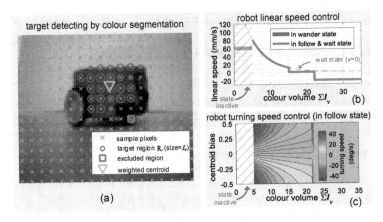

Fig. 2. Illustration of the target detection and following model. (a) The target detection, showing with the target robot staying 9 cm away. The red circle indicates the points belongs to the target, while the colour intensity is represented by the size of circle; (b) The robot's linear motion speed control behaviour. (c) The robot's rotation motion control by both target volume and centroid bias. In (b) and (c), when target volume is less than 5, the model is inactive which is overtaken by the CA model. (Color figure online)

procedure: (1) the identified target's colour volume D_{obj} (or the "mass" of target region), which is the zero-order moment of target region's value channel:

$$D_{obj} = \mu_{(0,0)}^{\mathbf{R}_r} = \sum_{i,j \in \mathbf{R}_r} \mathbf{I}_v(i,j) \tag{3}$$

and (2) the bias B_{obj} of the object's centroid against the image centre, which is the normalized horizontal first-order moment:

$$B_{obj} = \frac{\mu_{(1,0)}^{\mathbf{R}_r}}{\mu_{(0,0)}^{\mathbf{R}_r}} = \frac{\sum_{i,j \in \mathbf{R}_r} \mathbf{I}_v(i,j) \cdot (j - \frac{j_{max}}{2})}{\sum_{i,j \in \mathbf{R}_r} \mathbf{I}_v(i,j) \cdot j_{max}} \tag{4}$$

in which j is the horizontal index of a pixel, and j_{max} is the width of the image.

Motion control: The goal of motion control is to drive a robot towards the target to a region, where the target is in front of the follower with a certain distance. Specifically, this behaviour contains three phases:

P1: If the target emerges from far away, the robot accelerates to charge towards it, until the projected image has reached a certain size then becomes phase 3. The motion speed decrease exponentially during approach,
P2: If the target appears suddenly or becomes too close that contributes to a large region of image, which could happen when the target is moving, the robot retracts to leave enough space to enter P3.

P3: A *safe* zone is set when the size of target is satisfactory, then the robot enters the **wait** state. During this period, the robot only tunes its orientation slightly towards the centre of object but no displacement is made.

These phases can be described as functions of two group of variables: the robot's linear speed v and rotation speed ω controlled by the calculated metrics from image processing D_{obj} and B_{obj}, as illustrated in Figs. 2(b) and (c).

Notice that, by utilizing this visual navigation model alone, the initial catch up speed have to be greater than the target object, which could be another robot, in order to make this system converge [12].

2.3   The Hybrid Robot Control Strategy

The hybrid robot control strategy is based on the probabilistic approach: the finite state machine (FSM), since for such two visual models that generate conflicting behaviours, actions from only one model can be taken by the robot at a time according to its current situation, in order to ensure stability and consistency. In total, there are four behaviours generated by two models that need to be controlled, which are the **wander/turn** generated by the ELGMD model, and the **follow/wait** generated by the target following model. The structure and an example of robot behaviour are depicted in Fig. 3.

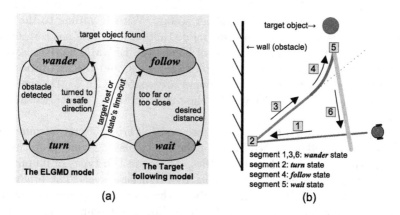

(a)   (b)

Fig. 3. The proposed finite state machine to incorporate the two visual models. (a) The illustration of proposed FSM, whose initial state is the *wander* state. (b) The illustration of a typical scenario controlled by this hybrid visual system.

In this FSM logic, the **wander** state is set to be the initial and default state, while the other states are restricted with conditions. In this way, suppose a robot inside an arena surrounded by walls, it would perform such a behaviour:

1. The robot wanders within the arena to establish autonomous navigation as long as possible, by trying to avoid collisions against walls or any other unidentified object by accepting motion commands from the ELGMD model.
2. When a target object is encountered, the robot enters the **follow** state to maintain certain distance behind. If the target object is stationary, the follower stops in the **wait** state. A timer is set once the motion commands from the target following model is activated. The activity last for a certain time t_w before it rotates to another direction, in order to return to the **wander** state to continue exploring the arena.

2.4   The Robot Platform and Experiment Arena Configuration

In this research, the micro robot platform *Colias IV* is utilized [17]. *Colias IV* is a micro ground robot occupying a circular footprint bearing the diameter of 4 cm. The motion is provided by two differential driven wheels. The *Colias IV* is equipped with a strong ARM-Cortex M4 micro-controller running at 180 MHz with embedded SRAM of 256 Kbytes, which ensures the required computation power for both proposed image processing tasks. The images used in the visual tasks are captured continuously by a tiny CMOS camera on-board with resolution configured at 99×72, 30 frames per second. With regarding to the frame rate, the system is working in real-time mode that each process has certain time allowance. In the total frame duration of 33 ms, the ELGMD need 7 ± 2 ms to process, the target finding model occupies 11 ± 1 ms, and the rest process use up to 1 ms. The robot in the experiments are coated with a 3D-printed red shell, in order to be distinguishable by the target following model, as shown in Fig. 4(a).

The experiments are conducted in an arena sizing 175×155 cm, surrounded by wall of 15 cm high and decorated by wall papers with black & white patterns. The floor of the arena is covered by black tiles. There are two types of illumination cast into this arena to separate two zones by brightness levels. The bright zone is irradiated by four LED spot lights above the arena, while the dark zone is lighted by ambient lamps of the whole room. Both light temperature has been tuned to 4000 K for consistency. Equipped with SORAA SNAP systems, the light beams produced by the LED have straight and sharp edges, leaving the transition border with thickness of almost 5 cm. Two red cue balls with diameter of $\phi = 48$ cm are fixed in the arena, serving as static stimuli. One cue ball is placed in centre of the bright zone and the other one in the dark zone.

Robots behaviour were recorded by overhead camera. Their trajectories are tracked and analysis by whycon system [18] by utilizing the circle markers on top of the robot [19], as shown in Fig. 4.

3   Experiments and Results

3.1   Individual Robot Performance Tests

The performance of hybrid visual model on the individual robot is tested accordingly. All tests in this subsection is done in the bright zone. Results of the data was illustrated in Fig. 5.

First, the stability of proposed visual models are tested by challenging the robot with static stimuli, as illustrated in Fig. 5(a).The following metrics are analysed and each scenario is repeated for 50 times:

1. the success rate of collision detection and avoidance η_c;
2. the Distance to Collision (DTC) of the ELGMD model for both the wall obstacle d_{tc}^{wall} and a green ball d_{tc}^{ball} for comparison, which is the distance between the robot and the obstacle when it turns away;
3. the ratio of switching models in a competitive scenario P_t, which is defined by: when two objects are placed in front of the robot, the ratio of switching from **wander** to **follow** and from **wander** to **turn**;
4. the distance where the robot stops in front of a red object $d_{att}(1)$, or two objects $d_{att}(2)$.

Results of these tests are concluded in Table 2. The results have demonstrated that the visual models work stable to realize their own goals. Moreover, it reveals that, for the visual target following model, the sensitivity and range increases as the target cluster expands, since a larger region of targets contributes to more obvious visual cues to be identified from far away.

Then, the variables of a robot deployed inside the arena are logged and analysed to show the dynamics of the visual models. For a typical scenario where the robot's trajectory is similar to Fig. 3(b), the data are plotted in Fig. 5(b). The recorded results show that the ELGMD has evoked a turning action when a collision is detected at around 4.5 s. Then at 8.3 s, a target object has been identified thus attracts the robot to charge towards it. The orientation is tuned continuously. Finally at 10.8 s, the robot stops behind the object. The series of states switching meet our expectations.

Table 2. The test results of individual robot behaviours

Name	η_c	d_{tc}^{wall}	d_{tc}^{ball}	P_t	$d_{att}(1)$	$d_{att}(2)$
Value	96%	41:9	8.3 ± 2.7 cm	6.1 ± 1.8 cm	14.1 ± 4.6 cm	21.6 ± 3.5 cm

3.2   Robot Aggregation Tests

Since the robots can join and leave the cluster at any time, one typical approach to study the stability of aggregation which is dynamically changing is to study the discriminative cluster size controlled by environment. By adjusting the time

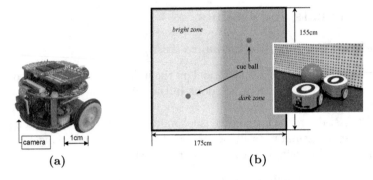

Fig. 4. The robot platform and experiment arena configuration. (a) The *Colias IV* robot deployed in this research. (b) The experiment arena set-up, showing the photo of a cue ball and *Colias IV* covered by 3D printed shells. (Color figure online)

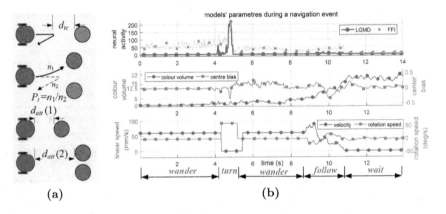

Fig. 5. The tests of hybrid model on individual robot. (a) Schematic of measuring the metrics of the hybrid model. The results are shown in Table 2; (b) One piece of the variables of a running robot during an arena test. The robot's trajectory is similar to which of Fig. 3(b). The robot's states are indicated at the bottom of plots.

that each robot remains in the cluster, i.e. the t_w, it is obvious that the longer time each robot stays inside a cluster, the larger size.

Being inspired by the aggregation behaviour found in honeybees and cockroaches [20], we employ the ambient illumination as the controlling factor, which is embodied by the camera's exposure intensity $I_e \in [0, 255]$, an integer automatically adjusted in the camera by a build-in histogram-based algorithm. Thus no additional sensors or communication methods are required. Its intention is to maintain the brightness level of captured images. Therefore, the value is tightly tuned by the view of the robot.

One test on the relationship between I_e and the robot's vision is illustrated in Fig. 6. The results showed that a worse illumination condition would be compensated by a higher exposure intensity, and vice versa. By using this mechanism, the I_e then tunes the robot's t_w by:

$$t_w = 70 \cdot \tanh(5 - \frac{I_e}{10}) + 20 \tag{5}$$

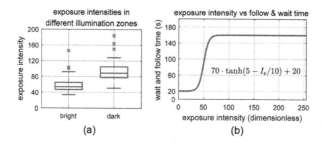

Fig. 6. (a) The tested exposure intensities under different illumination status; (b) The curve to tune wait time t_w by exposure intensity I_e.

The Aggregation Experiments: The experiments are conducted with different population sizes, given by 7, 11 and 15 robot agents respectively. Each experiment last for 15 min. At the beginning the robots were placed randomly inside. Aggregation number of both bright zone and dark zone were recorded and logged. Since there is no complete stop condition for any robot, the formation of groups are always fluctuating. This doesn't interfere us to analyse their properties by comparing the group sizes along time. For all three trials, we investigated the largest cluster in both zones along time. Records and results of the experiments are redrawn in Fig. 7.

Results: Result of the experiments show that robots tend to aggregate at darker zones rather than bright zones regardless of group sizes. For each group size, the dark zone always restrain robots for longer thus there are higher chances for more robots to meet and stop. On the other hand, the cluster in the bright zone vanishes quickly due to the short t_w. As a main factor that cause the environment selective behaviour, this dynamic behaviour is one of the common point in probabilistic aggregation. This behaviour is also found in the cockroaches inspired robots [20,21], whose chance to join or leave a cluster is affected by both the cluster size and environment darkness.

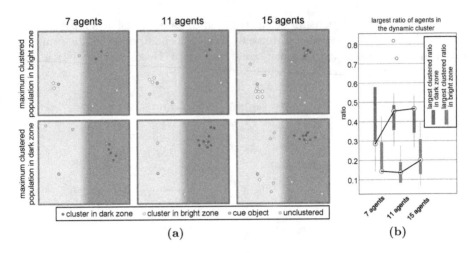

Fig. 7. The illustration of arena experiments results with multiple robots when the largest clustered population has reached in either zones (see scales in Fig. 4b); (b) The largest cluster population distributions in all experiment trials.

3.3 Discussion

Micro robots with strong vision ability have shown its potential in swarm robotics, but the approaches are not always straightforward. Being different from the utilization of conventional sensors such as sound, light or infra-red sensors [22], information from images is highly sensitive to the relative motion of target and viewing angle [23]. In earlier researches, some approaches use additional information [23] such as infra-red sensors to identify surrounding environment and prevent collisions, while some employ an omnidirectional camera [12] to expand viewing angle.

Although the vision has shown its potential in acting as the primary sensor to achieve multi-tasking [24], care should be taken at designing to get stable outputs for before motion controlling. Since the robots can only detect objects only in view, they cannot react to objects in other directions [5], thus aggregation is less likely to trigger than those robots equipped with omnidirectional sensors [22]. The cue balls act as static stimulus in the arena. They provide stable signals which help robots to gather around faster. As the population size gets larger, there is a higher chance that two robots run in a face to face trajectory, which causing a group without the help from cue balls.

4   Conclusion

The paper has discussed an approach to mediate two visual tasks from different categories and even with conflicting purposes in micro robots to serve multi-purpose tasks. This hybrid visual coordination model is described by a FSM that activates either model according to the robot's situation thus no interference is introduced to each other. The two visual models are a bio-inspired collision detection neural model, and a colour-sensitive target following algorithm. Both vision systems are computationally efficient and stable to be implementation on a microprocessor. The proposed hybrid model is realized on the micro robots named *Colias IV* to demonstrate the robustness.

Systematic experiments showed that the bio-inspired collision avoidance model ELGMD can cooperate with other visual models to achieve multi-tasks. The proposed hybrid visual model is feasible to be deployed in low-cost micro robots especially for educational purposes, which enables the collective behaviour study by micro robots utilizing normal camera as the only sensor.

In the future, study will be conducted on (1) to further investigate the aggregation behaviour evoked by visual inputs in swarm scenarios; (2) to improve the proposed target following models to achieve better precision and sensitivity.

Acknowledgement. This work is supported by EU-FP7-IRSES project HAZCEPT (318907) and Horizon2020 project STEP2DYNA.

References

1. Zufferey, J.-C., Floreano, D.: Toward 30-gram autonomous indoor aircraft: vision-based obstacle avoidance and altitude control. In: IEEE International Conference on Robotics and Automation (ICRA), pp. 2594–2599 (2005)
2. Meyer, H.G., Bertrand, O.J.N., Paskarbeit, J., Lindemann, J.P., Schneider, A., Egelhaaf, M.: A bio-inspired model for visual collision avoidance on a hexapod walking robot. In: Lepora, N.F.F., Mura, A., Mangan, M., Verschure, P.F.M.J.F.M.J., Desmulliez, M., Prescott, T.J.J. (eds.) Living Machines 2016. LNCS (LNAI), vol. 9793, pp. 167–178. Springer, Cham (2016). https://doi.org/10.1007/978-3-319-42417-0_16
3. Fu, Q., Hu, C., Liu, T., Yue, S.: Collision selective LGMDs neuron models research benefits from a vision-based autonomous micro robot. In: IEEE/RSJ International Conference on Intelligent Robots and Systems 2017, pp. 3996–4002 (2017)
4. Ardin, P., Peng, F., Mangan, M., Lagogiannis, K., Webb, B.: Using an insect mushroom body circuit to encode route memory in complex natural environments. PLoS Comput. Biol. **12**(2), e1004683 (2016)
5. Bagheri, Z., Cazzolato, B., Grainger, S., O'Carroll, D., Wiederman, S.: An autonomous robot inspired by insect neurophysiology pursues moving features in natural environments. J. Neural Eng. **14**(4), 046030 (2017)
6. Serres, J., Ruffier, F.: Optic flow-based robotics. Wiley Encyclopedia of Electrical and Electronics Engineering (2016)
7. Zufferey, J.-C., Floreano, D.: Fly-inspired visual steering of an ultralight indoor aircraft. IEEE Trans. Robot. **22**(1), 137–146 (2006)
8. Hu, C., Arvin, F., Xiong, C., Yue, S.: Bio-inspired embedded vision system for autonomous micro-robots: the LGMD case. IEEE Trans. Cogn. Dev. Syst. **9**(3), 241–254 (2017)
9. Rind, F.C., Simmons, P.J.: Seeing what is coming: building collision-sensitive neurones. Trends Neurosci. **22**(5), 215–220 (1999)

10. Bermudez i Badia, S., Pyk, P., Verschure, P.F.M.J.: A fly-locust based neuronal control system applied to an unmanned aerial vehicle: the invertebrate neuronal principles for course stabilization, altitude control and collision avoidance. Int. J. Robot. Res. **26**(7), 759–772 (2007)

11. Franceschini, N., Pichon, J.-M., Blanes, C.: From insect vision to robot vision. Phil. Trans. R. Soc. Lond. B **337**(1281), 283–294 (1992)

12. Das, A.K., Fierro, R., Kumar, V., Ostrowski, J.P., Spletzer, J., Taylor, C.J.: A vision-based formation control framework. IEEE Trans. Robot. Autom. **18**(5), 813–825 (2002)

13. Benavidez, P., Jamshidi, M.: Mobile robot navigation and target tracking system. In: 2011 6th International Conference on System of Systems Engineering (SoSE), pp. 299–304. IEEE (2011)

14. Arvin, F., Turgut, A.E., Bazyari, F., Arikan, K.B., Bellotto, N., Yue, S.: Cue-based aggregation with a mobile robot swarm: a novel fuzzy-based method. Adapt. Behav. **22**(3), 189–206 (2014)

15. Bermudez i Badia, S., Bernardet, U., Verschure, P.F.: Non-linear neuronal responses as an emergent property of afferent networks: a case study of the locust lobula giant movement detector. PLoS Comput. Biol. **6**(3), e1000701 (2010)

16. Loesdau, M., Chabrier, S., Gabillon, A.: Hue and saturation in the RGB color space. In: Elmoataz, A., Lezoray, O., Nouboud, F., Mammass, D. (eds.) ICISP 2014. LNCS, vol. 8509, pp. 203–212. Springer, Cham (2014). https://doi.org/10.1007/978-3-319-07998-1_23

17. Hu, C., Fu, Q., Yue, S., Colias IV: the affordable micro robot platform with bio-inspired vision. In: 19th Towards Autonomous Robotic Systems (TAROS) (2018)

18. Krajník, T., Nitsche, M., Faigl, J., Vaněk, P., Saska, M., Přeučil, L., Duckett, T., Mejail, M.: A practical multirobot localization system. J. Intell. Robot. Syst. **76**(3–4), 539–562 (2014)

19. Lightbody, P., Krajník, T., Hanheide, M.: A versatile high-performance visual fiducial marker detection system with scalable identity encoding. In: Proceedings of the Symposium on Applied Computing, pp. 276–282. ACM (2017)

20. Correll, N., Martinoli, A.: Modeling self-organized aggregation in a swarm of miniature robots. In: IEEE 2007 International Conference on Robotics and Automation Workshop on Collective Behaviors Inspired by Biological and Biochemical Systems, no. SWIS-CONF-2007-002 (2007)

21. Garnier, S., Gautrais, J., Asadpour, M., Jost, C., Theraulaz, G.: Self-organized aggregation triggers collective decision making in a group of cockroach-like robots. Adapt. Behav. **17**(2), 109–133 (2009)

22. Kernbach, S., Häbe, D., Kernbach, O., Thenius, R., Radspieler, G., Kimura, T., Schmickl, T.: Adaptive collective decision-making in limited robot swarms without communication. Int. J. Robot. Res. **32**(1), 35–55 (2013)

23. Gauci, M., Chen, J., Li, W., Dodd, T.J., Groß, R.: Self-organized aggregation without computation. Int. J. Robot. Res. **33**(8), 1145–1161 (2014)

24. Denuelle, A., Srinivasan, M.V.: Bio-inspired visual guidance: from insect homing to UAS navigation. In: 2015 IEEE International Conference on Robotics and Biomimetics (ROBIO), pp. 326–332. IEEE (2015)

Learning and Adaptation

Neural Control and Synaptic Plasticity for Adaptive Obstacle Avoidance of Autonomous Drones

Christian Koed Pedersen[1] and Poramate Manoonpong[1,2(⊠)]

[1] Embodied AI and Neurorobotics Lab, Centre for BioRobotics,
The Mærsk Mc-Kinney Møller Institute,
University of Southern Denmark, Odense M, Denmark
`chped13@student.sdu.dk, poma@mmmi.sdu.dk`
[2] Bio-inspired Robotics and Neural Engineering Lab,
School of Information Science and Technology,
Vidyasirimedhi Institute of Science and Technology, Rayong, Thailand
`http://ens-lab.sdu.dk/`

Abstract. Drones are used in an increasing number of applications including inspection, environment mapping, and search and rescue operations. During these missions, they might face complex environments with many obstacles, sharp corners, and deadlocks. Thus, an obstacle avoidance strategy that allows them to successfully navigate in such environments is needed. Different obstacle avoidance techniques have been developed. Most of them require complex sensors (like vision or a sensor array) and high computational power. In this study, we propose an alternative approach that uses two simple ultrasonic-based distance sensors and neural control with synaptic plasticity for adaptive obstacle avoidance. The neural control is based on a two-neuron recurrent network. Synaptic plasticity of the network is done by an online correlation-based learning rule with synaptic scaling. By doing so, we can effectively exploit changing neural dynamics in the network to generate different turning angles with short-term memory for a drone. As a result, the drone can fly around and adapt its turning angle for avoiding obstacles in different environments with a varying density of obstacles, narrow corners, and deadlocks. Consequently, it can successfully explore and navigate in the environments without collision. The neural controller was developed and evaluated using a physical simulation environment.

1 Introduction

The use of drones in various applications (including inspection, environment mapping, and search and rescue operations) has expanded in recent years [1–3]. The applications often take place outside in open areas away from obstacles and people, but as the technology advances the opportunity for flight in more complex areas arises. These are environments such as indoor or urban areas with many obstacles, sharp corners, and deadlocks. However, it does require the drone

© Springer Nature Switzerland AG 2018
P. Manoonpong et al. (Eds.): SAB 2018, LNAI 10994, pp. 177–188, 2018.
https://doi.org/10.1007/978-3-319-97628-0_15

to feature a collision avoidance strategy to enable safe flight. Often this strategy is achieved by using cameras and computer vision with algorithms (like, SURF or classifiers) to detect frontal objects and estimate a distance to them during flying [4,5]. The knowledge is then used to avoid obstacles before colliding them by moving to either side of the object. These vision-based algorithms sometimes cannot handle corner cases or scenarios where it is not a single enclosed entity in the way. To overcome this problem, other methods try to map the environment using algorithms (such as SLAM) and afterwards use the information for path planning to avoid obstacles [6]. This may require high computational power which can be difficult to implement on drones, especially when they are small in size. In this study, we propose an alternative approach inspired by [8] where they use simple two sensors and neural control with synaptic plasticity for adaptive obstacle avoidance of walking robots. Here we apply it to controlling autonomous drones. The neural control is based on a two-neuron recurrent network. Synaptic plasticity of the network is done by an online correlation-based learning rule with synaptic scaling. This neural-based control technique is a simple but effective way to enable a drone to navigate in complex environments with obstacles, narrow corners, and deadlocks. Due to low computational resource requirements for the neural algorithm implementation, this can open up for small indoor drone applications to fly in the environment with varying obstacle densities.

2   Materials and Methods

In this study, adaptive obstacle avoidance of an autonomous drone is achieved via a sensorimotor loop which involves neural dynamics, synaptic plasticity, sensory feedback, and a physical drone body (Fig. 1). Neural dynamics and plasticity are embedded in a neural control network (Fig. 2). Sensory feedback from the left and right obstacle detection sensors of the drone is processed through the network. The network outputs are used for pitch and yaw control of the drone. This results in an autonomous drone system that can navigate in a complex environment without collision. The drone is simulated in a V-REP and the communication between the simulated drone and the neural control network is based on the Robot Operating System (ROS).

2.1   System Overview

The drone system is based on two main entities (see Fig. 1). It consists of a V-REP [7] simulation and a ROS-based neural control system.

Here we use a yaw rate and attitude roll and pitch parameters as control inputs to the drone. A positive value of the yaw rate will drive the drone to turn clockwise around its own axis. A negative value will make it turn counter clockwise. Two ultrasonic-based distance sensors are implemented at the front part of the drone for obstacle detection. Each sensor has a beam angle of $10°$ and a range of 0.5 m. The sensors are rotated $35°$ and positioned 5 cm with no pitch from the forward pointing axis of the drone. Height control is not the main focus here; therefore, the height setpoint is kept fixed at a certain height in all experiments. The neural controller is based on ROS and written in C++. The ROS interface from the simulation enables the controller to get sensory information and send motor commands to the drone simulation. The controller implements the neural control network and synaptic plasticity described in the following sections.

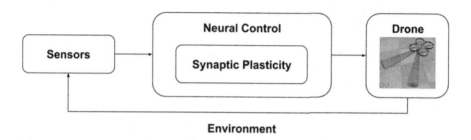

Fig. 1. System overview of adaptive embodied neural closed-loop control for autonomous obstacle avoidance and navigation behaviours of a drone.

2.2 Neural Control for Obstacle Avoidance of Autonomous Drones

The neural control for obstacle avoidance of an autonomous drone is derived from a neural network shown in Fig. 2. The basic principle of this neural control approach is inspired by Braitenberg vehicle [9]. Here it is adapted to gain additional dynamics for a drone platform rather than a two wheeled robot. All neurons of the network are modeled as discrete-time non-spiking neurons. The activity a_i of each neuron develops according to:

$$a_i(t) = \sum_{j=1}^{n} W_{ij}o_j(t-1) + B_i, i = 1, ..., n \qquad (1)$$

where n denotes the number of units, B_i an internal bias term or a sensory input to neuron i, W_{ij} the synaptic strength of the connection from neuron j to neuron i. The output o_i of all neurons of the network is calculated by using a hyperbolic tangent (tanh) transfer function, i.e., $o_i = tanh(a_i), \in [-1, 1]$, except for the two neurons (N6 and N7, see Fig. 2) using a sigmoid transfer function.

The core of the network consists of the two recurrent input neurons (N1 and N2, see Fig. 2). They each receive obstacle detection signals from two ultrasonic-based distance sensors, installed at the front part of the drone. The range of each sensor is adjusted to 50 cm. Before feeding the raw sensory signals to the network, we map them to the range of $[-1, 1]$ where -1 means no obstacle in the range and 1 means that an obstacle is near (about 20 cm distance). The output signals ($O_{1,2} \in [-1, 1]$) of the two neurons are transmitted to the output neuron (N3). The output of N3 is then mapped to a yaw command which is finally sent to the drone.

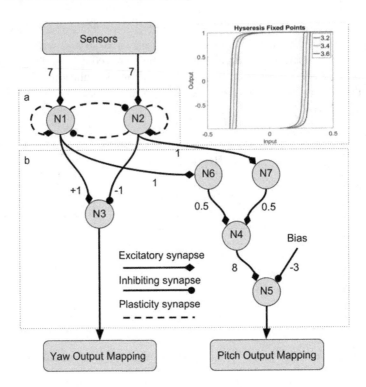

Fig. 2. The neural control network for obstacle avoidance of drones. (a) The two recurrent input neurons with plasticity synapses. (b) The drone specific network used for yaw and pitch control. (c) An example of input-output relation of a recurrent neuron (i.e., N1) of the network. The neuron shows different hysteresis loops at different self-connection weights.

According to this setup, the yaw command will be a positive value if the drone detects an obstacle on the left and a negative value if it detects an obstacle on the right.

The self-connections (W_{11}, W_{22}) can lead to a hysteresis effect in their neural activations. The hysteresis assures that the drone can perform a continuous turning behavior although the sensors do not detect an obstacle anymore. Note, the strength of the self-connections (>1.0) defines the hysteresis width which determines the turning angle in front of the obstacles for avoiding them, i.e., the larger the hysteresis interval, the larger the turning angle. Additionally this hysteresis effect can reduce the amount of a wall following behavior that a network without the self-connections or Braitenberg vehicle-based control will generate.

Besides the self-connections, the neurons are mutually connected by two inhibitory synapses (W_{12}, W_{21}). This forms a so-called even loop [10]. In an even loop, in general, only one neuron at a time is able to produce a positive output, while the other one has a negative output, and vice versa. This guarantees the optimal functionality for avoiding obstacles and escaping from getting stuck in corner cases. In other words, in a corner the neural system can remember the first obstacle for a certain duration and executes the corresponding behavior (turn) and, in addition, it does not trigger the behavior induced by the second obstacle which would be a turning in the opposite direction.

However, since this network only affects yaw, the drone will always continue to move forward depending on the pitch setpoint given. In case there is not enough space to perform a forward-moving turn for avoiding an obstacle, the drone might collide the obstacle. To overcome the problem, the second control network (N4–N7) was developed (Fig. 2). It is applied for pitch control and works by first changing the activation range of the two input neurons from $[-1, 1]$ to $[0, 1]$ using simple neurons with a sigmoid transfer function. Next, it finds the average by having synapses with weights of 0.5 from each sigmoid neuron (N6, N7) to a hidden neuron (N4). The output of N4 represents the distance and presence of nearby objects with a range of $[0, 1]$. A bias term with inhibitory synapse is introduced to the final output neuron (N5), with the purpose of driving the output towards negative values when no objects are detected. Finally the resulting output from neuron N5 is used as a scaling of the pitch setpoint. This leads to reduction of forward motion when objects become closer and at a certain point also backward motion when objects are very close to the drone.

The ranges for yaw and pitch setpoints are arbitrary in the simulation and do not represent ordinary units such as m/s or similar, and thus the final yaw and pitch outputs need to be mapped appropriately. The weights of the recurrent and inhibitory synapses between the two input neurons, are responsible for the adaptive obstacle avoidance behavior of the drone. In this study, the self-connections are initially set to a positive value (e.g., 0.5) and the mutual connections between N1 and N2 are initially set to a negative value (e.g., -0.5). To obtain proper connection weights for avoiding obstacles, corners, and deadlocks in different environments, we apply synaptic plasticity mechanisms described in the following section. The other weights of the network are statically set as depicted in Fig. 2.

2.3 Synaptic Plasticity for Adaptive Obstacle Avoidance

To achieve adaptive obstacle avoidance of an autonomous drone (i.e., online adaption of its turning angle for avoiding obstacles in different environments with a varying density of objects, narrow corners, and deadlocks), we use here correlation-based learning [11] to modify the four synaptic weights ($W_{11,22}$ and $W_{12,21}$) of the two-neuron recurrent network, part of the neural control network (see Fig. 2). This learning rule is based on three factors: The output activity $O_i(t)$ ($i \in 1, 2$) of the neurons ($N_{1,2}$) at the time step t, the output activity $O_i(t-1)$ of the neurons ($N_{1,2}$) at the previous time step (t − 1), and a reflex signal $R_i(t)$ computed from the sensory input I_i as described in Eq. 2.

$$R_i(t) = \begin{cases} 1, & \text{if } I_i(t) > -0.5, \\ 0, & \text{if } I_i(t) \le -0.5. \end{cases} \tag{2}$$

The reflex signal is used to control the learning process which will be activated as soon as the drone detects an obstacle at close range (about 0.3 m). Additionally, we also employ synaptic scaling [12] to avoid instability of the synaptic weights. To assure that the synaptic weights do not change their sign in order to maintain the hysteresis effects, for the learning rule, we map the outputs $O_i \in [-1, 1]$ to the positive interval $v_i \in [0, 1]$ as shown in Eq. 3.

$$v_i = \frac{O_i + 1}{2}. \tag{3}$$

The self-connections $(W_{11,22})$ are governed by Eqs. (2 and 3), respectively.

$$W_{11}(t+1) = W_{11}(t) + \mu_r \cdot v_1(t-1) \cdot v_1(t) \cdot R_1(t) + \gamma(k - v_1(t)) \cdot W_{11}(t)^2, \tag{4}$$

$$W_{22}(t+1) = W_{22}(t) + \mu_r \cdot v_2(t-1) \cdot v_2(t) \cdot R_2(t) + \gamma(k - v_2(t)) \cdot W_{22}(t)^2. \tag{5}$$

μ_r is the learning rate that changes the time scale of the learning process. It is here set to 0.01. γ is the forgetting rate that similarly changes the time scale of forgetting/synaptic scaling. It is set 0.0003. k is the offset that enables a constant reduction of the weights when no obstacle is detected. It is set to -0.01.

The two inhibitory connections $(W_{12,21})$ are modified in a similar fashion, but are changed equally to maintain symmetry; thereby forming the even loop [10]. This is done by introducing temporary variables $(q_{1,2})$ that calculate from them the average inhibitory connections. These parameters are updated as follows:

$$q_1(t+1) = \mu_q \cdot v_1(t-1) \cdot v_1(t) \cdot R_1(t) + \gamma(k - v_1(t)) \cdot q_1(t)^2, \tag{6}$$

$$q_2(t+1) = \mu_q \cdot v_2(t-1) \cdot v_2(t) \cdot R_2(t) + \gamma(k - v_2(t)) \cdot q_2(t)^2, \tag{7}$$

$$W_{12,21}(t+1) = W_{12,21}(t) + 1/2 \cdot (q_1(t+1) + q_2(t+1)). \tag{8}$$

μ_q is the learning rate and set to 0.015. γ and k are the forgetting rate and the offset and they are set to the values described above. Note that all meta parameters (i.e., the learning and forgetting rates and the offset) are empirically selected.

In principle, this learning mechanism changes the synaptic weights of the two-neuron recurrent network in a way that the sensory inputs and the weights will drive the neural output to reach and stay at the upper fixed point $(\approx+1)$ of the hysteresis loop (see Fig. 2c) while the drone is trying to escape from a corner or a deadlock. As a consequence, the drone will escape from the situation by performing a very large turning angle. Once it has escaped or does not detect an obstacle any more, the second part of the mechanism (synaptic scaling) will reduce the synaptic weights such that the neural output returns to the lower fixed point (≈-1) of the hysteresis loop (see Fig. 2c); thereby the drone will stop turning and continue to fly forward. In other words, the interaction of correlation-based learning and scaling moves the neural system between two fixed point states (i.e., hysteresis effects). Note that the used learning mechanism is independent of the initial weight values [12].

3   Experiments and Results

The neural controller described in this study was tested in different environments with and without the use of synaptic plasticity. Three environments with a varying density of objects (see Fig. 3) were created. The first and second environments (Figs. 3a and b) have sparse and dense random obstacles, respectively, while the third one has obstacles that were placed to form corners and deadlocks (Fig. 3c).

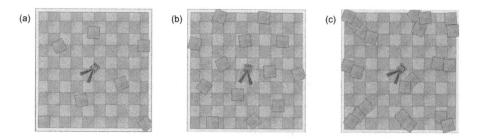

Fig. 3. Simulated environments. (a) Low density of obstacles. (b) High density obstacles. (c) High density of obstacles with corners and deadlocks.

The tests were conducted by setting a constant forward motion with a small random noise to the yaw control to enable an exploration behavior. Each test was run for approximately 30 min. Otherwise it was manually stopped if the drone got stuck. Figure 4 shows the simulation results of the drone using synaptic plasticity in the three environments. We initially set the synaptic weights of the recurrent network to $W_{11,22} = 0.5$ and $W_{12,21} = -0.5$ and changed them through the online learning rule described above. It can be seen that the drone could fly in all environments without getting stuck.

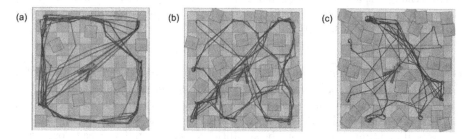

Fig. 4. (a), (b), (c) The trajectories of the drone in the three different environments or maps. Notice that no matter the density of objects, the drone is capable of moving around in all areas of the maps.

Figure 5 shows synaptic weight changes during the experiments. It can be seen that, in the environment with low dense obstacles (Fig. 4a), the weights

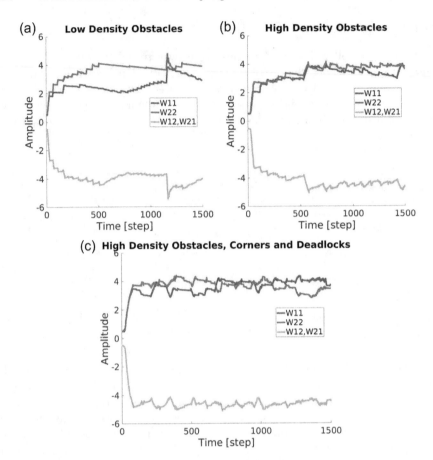

Fig. 5. The progress of weights during tests using synaptic plasticity. The weights change differently depending on the environment. (a) In the sparse environment, the weights changes slowly, up until a point (around 1200 time steps) where the drone enters one of the few narrow corner cases. (b), (c) When the density is increased, the weights start to change more rapidly.

changed slowly. This is because the drone did not frequently detect obstacles during the flight (Fig. 5a). In contrast, the weights increased quite fast (Fig. 5b) in the environment with high dense obstacles (Fig. 4b) and very fast (Fig. 5c) in the environment with high dense obstacles, corners, and deadlocks (Fig. 4c). In all cases, if no input was present, the synaptic weights started to decay slowly due to synaptic scaling which prevents the divergence of the weights. Table 1 shows three sets of average weights obtained from Fig. 5. These weights were used as the fixed weights ($W_{11,22,12,21}$) of the recurrent network to test the drone without online adaption.

Figure 6 shows the simulation results of the drone with the three sets of the fixed average weights (Table 1) without online adaption in the corresponding

Table 1. Weight averages

Environment	$W_{11,22}$	$W_{21,12}$
Low density of obstacles	3.1707	−3.9627
High density of obstacles	3.3689	−4.2286
High density of obstacles corners and deadlocks	3.6105	−4.5302

environments. In the environments with high dense obstacles, corners, or/and deadlocks, the drone could not fully explore the areas. It basically got stuck in a corner (Fig. 6b) or a deadlock (Fig. 6c). Only in the environment with low dense obstacles, the drone can fly without getting stuck (Fig. 6a).

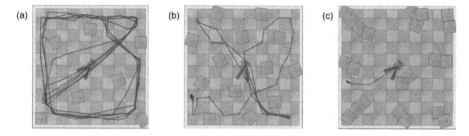

Fig. 6. (a), (b), (c) The trajectories of the drone without online adaptation in the three different environments. Notice that in the environment with low density of obstacles (a), the drone is capable of getting to most areas and avoids most obstacles. In the environment with high density obstacles (b), it ends up at a corner and cannot get away. In the environment with high density of obstacles, corners, and deadlocks (c), it almost immediately hits a corner where it gets stuck.

A quantitative test was made to further verify the improvements of the synaptic plasticity. The test was conducted by simulating each environment, with and without the synaptic plasticity. A run was deemed successful when the drone was able to fly for 5 min without either getting stuck or crashing. A total of 5 runs was done for each test and the results are showed in Table 2.

The results of the experiments show that synaptic plasticity adds adaptation to a simple recurrent network, and thus enables the drone to fully explore in the environments with a varying density of obstacles, narrow corners, and deadlocks. It also allows the drone to autonomously avoid obstacles and escape from a corner and a deadlock. This effect can be compared when plasticity (see Figs. 4b, c) and non-plasticity (see Figs. 6b, c) are used. When observing the weights during flying in the sparse obstacle environment (Fig. 5a), there are small incremental changes for a long period. This is expected to happen when the drone reaches corners or obstacles where it only needs a larger turning angle to avoid getting stuck. However, at around the 1200 time steps (Fig. 5a) there is a large spike

Table 2. Results of quantitative tests run on each environment. The runs are deemed successful if the drone is able to fly for 5 min without crashing or getting stuck.

Environment	Low density	High density	High density, sharp corners and deadlocks
Ratio of successful runs with synaptic plasticity	5/5	4/5	4/5
Ratio of successful runs without synaptic plasticity	5/5	3/5	0/5

in the weight values indicating that it has reached a case where a large turning angle is required to escape from the situation. Without this adaptation, a large turning angle would be required at all times even though it is unnecessary and excessive in most cases. The results indicates that this method could very well be used for collision avoidance.

4　Discussion

Obstacle avoidance is an important basic function for autonomous navigation of drones. According to this, different obstacle avoidance techniques have been developed [4–6,16,17]. Many of them use visual cameras to extract information regarding obstacles. The information is used to find a suitable path to fly around the obstacles. Another way is to use Braitenberg's approach [9] which reactively controls an agent based on the activations of sensory inputs. For this approach, the agent will turn as long as it detects an obstacle. At a corner or deadlock, it might switch between turning left and right several times or it sometimes gets stuck. To deal with such a problem, Toutounji and Pasemann [18] proposed self-regulating neurons with short-term plasticity [19]. This allows an agent to avoid sharp corners. However, in this approach, different to our control proposed here, it requires multiple distance sensors for successful obstacle avoidance and the activity states have to be predefined.

In contrast to the previous approaches, we introduce neural control with synaptic plasticity which results in an adaptive obstacle avoidance strategy by using only two distance sensors. It is based on a two-neuron recurrent network with online correlation-based learning which adapts the synaptic weights of the network and thus changes the neural dynamics [13].

Recurrent networks have been proven useful in robot control [14,15] and in this case enables the drone to successfully navigate in complex environments by avoiding collision and escaping from deadlocks or sharp corners. The simple recurrent network does not rely on global or memorized information, but, instead, learns and adapts to each situation as its arise.

In the future, we plan to transfer the neural control approach to a real drone. Accidental collisions can lead to damage to the drone. Thus, to ensure the safety in case the proposed control fails, propeller protection or prop guards will be installed on the drone, like the Lumenier Danaus drone[1]. This will allow for minor collisions where the control will still be able to learn and react before serious damage will occur.

5 Conclusion

In this study, we introduced neural control and synaptic plasticity for adaptive obstacle avoidance of an autonomous drone. The V-REP simulator was used to simulate a drone and different environments as well as to evaluate the performance of the developed controller and demonstrate the obstacle avoidance behavior. The core control mechanism is based on a two-neuron recurrent network. The network receives sensory inputs and translates them into proper motor commands through post-processing neural units for pitch and yaw control. Online correlation-based learning with synaptic scaling was employed for synaptic plasticity of the recurrent network. In principle, this online learning mechanism can increase or decrease the weights with respect to the interaction of the drone with its environment; thereby changing neural dynamics in the network (e.g., various hysteresis effects). This changing neural dynamics can be utilized for generating different turning angles with short-term memory when facing to different obstacles, corners, and deadlocks. This results in adaptively avoiding obstacles and escaping from corners/deadlocks. Furthermore, this also enables the drone to successfully explore and navigate in cluttered unknown environments. In the future, we will transfer the developed neural control approach to a real drone in order to test the adaptive obstacle avoidance behavior in a real environment.

Acknowledgments. This research was supported partly by Center for BioRobotics (CBR) at the University of Southern Denmark (SDU) and Startup Grant-IST Flagship research of Vidyasirimedhi Institute of Science & Technology (VISTEC).

References

1. Ashour, R., Taha, T., Mohamed, F.: Site inspection drone: a solution for inspecting and regulating construction sites. In: Proceedings of the IEEE 59th International Midwest Symposium on Circuits and Systems, pp. 1–4 (2016)
2. Sanfourche, M., Le Saux, B., Plyer, A., Le Besnerais, G.: Environment mapping & interpretation by drone. In: Joint Urban Remote Sensing Event, pp. 1–4 (2015)
3. Pobkrut, T., Eamsa-ard, T., Kerdcharoen, T.: Sensor drone for aerial odor mapping for agriculture and security services. In: Proceedings of the 13th International Conference on Electrical Engineering/Electronics, Computer, Telecommunications and Information Technology, pp. 1–5 (2016)
4. Mori, T., Scherer, S.: First results in detecting and avoiding frontal obstacles from a monocular camera for micro unmanned aerial vehicles. In: Proceedings of the IEEE International Conference on Robotics and Automation, pp. 1750–1757 (2013)

---

[1] http://www.lumenier.com/products/multirotors/danaus.

5. Sedaghat-Pisheh, H., Rivera, A.R., Biaz, S., Chapman, R.: Collision avoidance algorithms for unmanned aerial vehicles using computer vision. J. Comput. Sci. Coll. **33**, 191–197 (2017)
6. Magree, D., Mooney, J.G., Johnson, E.N.: Monocular visual mapping for obstacle avoidance on UAVs. In: Proceedings of the International Conference on Unmanned Aircraft Systems, pp. 471–479 (2013)
7. Rohmer, E., Singh, S.P.N., Freese, M.: V-REP: a versatile and scalable robot simulation framework. In: Proceedings of the IEEE International Conference on Intelligent Robots and Systems, pp. 1321–1326 (2013)
8. Grinke, E., Tetzlaff, C., Wörgötter, F., Manoonpong, P.: Synaptic plasticity in a recurrent neural network for versatile and adaptive behaviors of a walking robot. Front. Neurorobot. **9**, 11 (2015)
9. Braitenberg, V.: Vehicles: Experiments in Synthetic Psychology (1986)
10. Pasemann, F.: Discrete dynamics of two neuron networks. Open Syst. Inf. Dyn. **2**, 49–66 (1993)
11. Kolodziejski, C., Porr, B., Wörgötter, F.: Mathematical properties of neuronal TD-rules and differential Hebbian learning: a comparison. Biol. Cybern. **98**, 259–272 (2008)
12. Tetzlaff, C., Kolodziejski, C., Timme, M., Wörgötter, F.: Analysis of synaptic scaling in combination with hebbian plasticity in several simple networks. Front. Comput. Neurosci. **6**, 36 (2012)
13. Neves, G., Cooke, S.F., Bliss, T.V.P.I.: Synaptic plasticity, memory and the hippocampus: A neural network approach to causality. Nat. Rev. Neurosci. **9**, 65–75 (2008)
14. Hülse, M., Pasemann, F.: Dynamical neural Schmitt trigger for robot control. In: Dorronsoro, J.R. (ed.) ICANN 2002. LNCS, vol. 2415, pp. 783–788. Springer, Heidelberg (2002). https://doi.org/10.1007/3-540-46084-5_127
15. Pasemann, F., Huelse, M., Zahedi, K.: Evolved neurodynamics for robot control. In: European Symposium on Artificial Neural Networks, pp. 439–444 (2003)
16. Zufferey, J.-C., Floreano, D.: Fly-inspired visual steering of an ultralight indoor aircraft. In: Proceedings of the Transactions on Robotics, pp. 137–146 (2006)
17. Franceschini, N., Ruffier, F., Serres, J., Viollet, S.: Optic flow based visual guidance: from flying insects to miniature aerial vehicles. INTECH Open Access Publisher (2009)
18. Toutounji, H., Pasemann, F.: Behavior control in the sensorimotor loop with short-term synaptic dynamics induced by self-regulating neurons. Front. Neurorobot. **8**, 19 (2014)
19. Zahedi, K., Pasemann, F.: Adaptive behavior control with self-regulating neurons. In: Lungarella, M., Iida, F., Bongard, J., Pfeifer, R. (eds.) 50 Years of Artificial Intelligence. LNCS (LNAI), vol. 4850, pp. 196–205. Springer, Heidelberg (2007). https://doi.org/10.1007/978-3-540-77296-5_19

Deep Feedback Learning

Bernd Porr[(✉)] and Paul Miller

Glasgow Neuro LTD, Glasgow, UK
{bernd,paul}@glasgowneuro.tech

Abstract. An agent acting in an environment aims to minimise uncertainties so that being attacked can be predicted, and rewards are not only found by chance. These events define an error signal which can be used to improve performance. In this paper we present a new algorithm where an error signal from a reflex trains a novel deep network: the error is propagated *forwards* through the network from its input to its output, in order to generate pro-active actions. We demonstrate the algorithm in two scenarios: a 1st-person shooter game and a driving car scenario, where in both cases the network develops strategies to become pro-active.

1 Introduction

When an agent acts in its environment, its actions in turn will change its sensor inputs which in turn will cause new actions – in short, the agent acts in a closed-loop system. In the simplest case this is a reactive system where the agent encounters threats or rewards and acts accordingly. Both threats and rewards are unpredictable events where the agent must respond as quickly as possible. These are often called "disturbances" or "perturbations" because they force the agent to act at an unpredictable moment in time. Importantly, these occur at the *input* of the agent and, thus, a closed loop system performs "input control" – in contrast to a pattern recognition system which performs "output control" [9].

The above scenario has a major drawback in that the reactions are always too late. It would be safer for an agent to learn to predict these disturbances from other cues, which introduces the concept of adaptive behaviour. At the moment of the disturbance it's already too late - the tiger has already attacked, or the food has been found by pure coincidence. However, one can then look back in time and determine which input signals could have predicted this unexpected threat or reward [10,14,19].

Adaptive controllers for this kind of learning are either correlation-based [11,17] or state space-based [14,18]. The state space-based ones can become very powerful by utilising a deep learning network. However, their main drawback is that they are slow and the state space approach limits their applicability to real world problems. In addition, the deep learning architecture requires backprop, which contradicts the feedforward nature of biologically-realistic networks [1], and can only be justified in special cases, e.g. by precisely tuned gating and/or simple feedback circuits [4]. On the other hand correlation based-methods can

© Springer Nature Switzerland AG 2018
P. Manoonpong et al. (Eds.): SAB 2018, LNAI 10994, pp. 189–200, 2018.
https://doi.org/10.1007/978-3-319-97628-0_16

be very fast, but so far it has been difficult to use them on deep networks. These correlation based methods use "input control", which means that the error signal is fed into the network at its inputs. This is much more compatible with biology, which requires a network that propagates errors in a feedforward fashion.

How could an error be propagated through a deep network in a forward fashion? Here we take inspiration from how errors are transmitted in backprop. In essence the hidden layers receive a *weighted sum* of the errors from the previous layers, which makes intuitive sense: the strongest weights contribute the most to the error. Imagine we take a similar approach using forward propagation: the error, introduced at the input, will have its strongest impact via the largest weights. Therefore we propagate the error in a forward fashion in the same way as in backprop. Eventually the agent will make an action and this will feedback to the input of the agent, which in turn then corrects the weights in all layers.

In this paper we present a novel deep network algorithm for closed loop systems which we call "Deep Feedback Learning" (DFL). This algorithm marries the ideas for correlation based closed loop learning with those from deep learning, by turning deep learning from a backprop-based algorithm to a forward-prop one. We demonstrate this with two scenarios: a driving task and a 1st-person shooter.

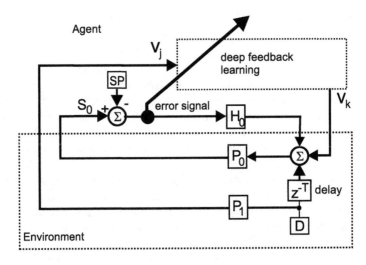

Fig. 1. A closed loop system with a setpoint SP, transfer function H and the environment P_0 and P_1 which needs to work against unpredictable disturbances D.

2   Closed Loop Learning

Deep feedback learning operates in a closed loop scenario. Before we describe the algorithm, we need to place it in a closed loop context. Figure 1 shows the entire closed loop system with the deep feedback learning as a black box for now. The idea is that we have a fixed closed loop which is able to fend off disturbances, such as an unexpected bend on a road or the sudden appearance of an enemy. This fixed loop then takes appropriate action to solve this disturbance, e.g. correcting a car's steering, or aiming towards and shooting an enemy. In formal terms we have a setpoint SP which compares the input of the organism to a desired input. If the input deviates from the setpoint an action is generated with the transfer function H_0. This action then eliminates the disturbance D and arrives via the environmental transfer function P_0 at the input again; thus, the loop is closed. However, we are not so much interested in the design of the closed loop but rather that it generates an *error signal*. This error signal is non-zero if a disturbance has happened, and can be used to tune our deep feedback learning network.

The deep feedback learning network receives additional inputs which are able to predict the disturbance, and thus prevent the trigger of the feedback loop. These additional inputs are provided via the transfer function P_1 and represent the disturbance in a filtered form. For example a video camera can provide images of the road ahead, or of an enemy. Deep feedback learning has the task to take the error signal, and tune its network to generate actions to minimise the error. In the context of the driving scenario the network output would be a steering action which would help to improve or replace the rigid feedback loop control. In the next section we describe the deep feedback learning and how this can compute the appropriate output. Note that deep feedback learning only knows if it has been successful once its actions have travelled through the environment. Thus learning evaluates if certain inputs to its network can be used to generate appropriate actions and these are then slowly transformed into actions. For that reason the error signal is propagated in a *forward* fashion through the network.

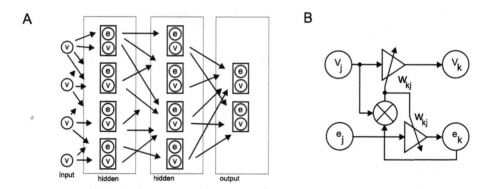

Fig. 2. (A) Network overview. With the exception of the input layer, every neuron is a composite cell with an activation v and an error term e. These are propagated through the network in a weighted fashion in parallel. (B) Computation in a single composite cell. The presynaptic activities v_j and error signals e_j are used to perform correlation based learning and change the weight w_{kj} which weights for both activity and error propagation towards the next layer.

3   Deep Feedback Learning

We define a network with an input layer, arbitrary hidden layers and an output layer which can all have different numbers of neurons (see Fig. 2A). In contrast to traditional networks, every layer (except for the input layer) consists of two summation nodes: the actual activity and an error signal. These are processed in two parallel streams.

Let us first focus on the network activity. We define a multi-layered network, where every neuron is a standard computational unit that calculates a weighted sum v_k of its inputs v_j and then applies an activation function $\Theta(v) = \tanh(v)$:

$$v_k = \Theta \left(\sum_j w_{jk} v_j \right) \tag{1}$$

where the activity flows from neurons in layer v_j to neurons in layer v_k multiplied by the weights w_{jk}, and this is then repeated in the next layer. This means that for a network with N layers we have $N - 1$ sets of weights w_{jk}.

The weight changes are then updated in a semi-Hebbian fashion:

$$w_{jk}(t+1) = w_{jk}(t) + \gamma v_j(t) e_k(t) \tag{2}$$

where $v_j(t)$ is the presynaptic activity and $e_k(t)$ is an error signal attached to the postsynaptic neuron, so the correlation is calculated between input signals and the error signal. The learning rate is γ.

We now describe the error signal propagation. As outlined above, the error signal emerges from the feedback loop, and is injected into the network at its 1st hidden layer as the "postsynaptic" activity. The weight change for this 1st layer can then be calculated directly with Eq. 2 by setting e_k to the error signal of the feedback loop (see Fig. 1).

For the deeper layers, the error signal is computed as a weighted sum of the error signals from the previous layer:

$$e_k = \frac{\left(\sum_j w_{jk} e_j \right) \Theta'(v_k)}{\frac{\sum_j |w_{jk}|}{\sum_j 1}} \tag{3}$$

where the $\Theta'(v) = 1 - v^2$ is the derivative of the activation function $\Theta(v)$: this limits learning when the unit approaches saturation, as in backprop. The norm guarantees that the error propagates through all layers and does not vanish from layer to layer due to small weights.

A simplified data flow diagram of the learning performed in each layer is shown in Fig. 2B, where we see the two processing streams: both the activity v_j and error signal e_j are weighted by w_{jk}, and summed separately in the next layer. Remember that for the 1st hidden layer, the error is just the error signal directly injected from the feedback loop. The diagram omits the error normalisation, scaling and activation derivative in order to focus on the main point that learning happens between the error signal and the presynaptic activity.

Learning is then performed in three steps: first the activity is propagated through the network, then the error signal is propagated via the same mechanism, and finally the weights are adjusted. Thus, both the error signal and the activity is propagated in a forward fashion. Learning itself can be interpreted as heterosynaptic for the activity v_j and Hebbian for the error e_j.

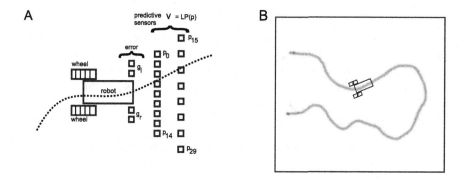

Fig. 3. (A) Robot setup. (B) The line following scenario used for the simulations.

4   Line Follower

In order to demonstrate DFL we need a simple closed loop scenario which can be improved with the help of DFL. Figure 3 shows a simple line following robot (Fig. 3A) which has two wheels whose speed is controlled by the fixed feedback control and DFL. The robot has the task of following the line depicted in Fig. 3B to the end, where it reverses and drives back, and so on. The robot is simulated with an updated version of ENKI for QT5 (https://github.com/berndporr/enki). The line follower is using just the ground sensors of the robot to create the error signal:

$$\text{error} = (g_{l_1} + 2g_{l_2}) - (g_{r_1} + 2g_{r_2}) \tag{4}$$

which directly creates a steering reaction from the fixed feedback loop by controlling the speed of the wheels. Each of the 30 predictive signals p_m from the two rows of ground sensors is split into ten 2nd-order lowpass filters with impulse responses lasting from two timesteps to 30 timesteps, and then all feed into the DFL network as predictive inputs v_j, giving 300 inputs in total. There were two hidden layers with nine and six neurons; six output neurons. All layers were fully connected. Learning was always on and the learning rate was 0.0001.

The output layer of our deep feedback learner consists of six neurons with activations v_k $(k = 0 \ldots 5)$ - these can be seen as soft decision-making units where three of them determine the change of speed of the right wheel, and three the left wheel. This leads to the final formulas for the motor outputs:

$$\text{leftSpeed} = s_0 + \underbrace{g\,\text{error} + (50v_0 + 10v_1 + 2v_2)}_{v_l} \tag{5}$$

$$\text{rightSpeed} = s_0 - \underbrace{g\,\text{error} + (50v_3 + 10v_4 + 2v_5)}_{v_r} \tag{6}$$

where $v_0, \ldots, v_5$ are the 6 outputs from the DFL network. Note that neither inputs nor outputs are organised in a topographically meaningful way. The network must discover from the error signals which sensor inputs v_k will eventually lead to appropriate steering actions. At the start the network is initialised with random values.

As performance criterion we use the squared average of the error from Eq. 4:

$$\text{error}_{\text{avg}} = \text{error}_{\text{avg}} + 0.001(\text{error} - \text{error}_{\text{avg}}) \tag{7}$$

$$\text{error}_{\text{sq}} = \text{error}^2_{\text{avg}} \tag{8}$$

As learning tries to minimise the average error, $\text{error}_{\text{avg}}$ should reach zero in an ideal scenario. Realistically, driving will never be 100% perfect, and $\text{error}_{\text{avg}}$ will stabilise at small values.

4.1   Results

Figure 4 shows the results of a simulation run. In the panels (A) and (D) we see the trajectory of the agent over the course of 11400 time steps of learning. While in (A) the agent clearly just follows the reflex reaction, leading to large deviations from the track, in (D) the agent closely follows the track and the deviation is minimal – learning has been successful. In (B) we see the network learning the steering actions that keep the agent closely on track. It can be seen that the network clearly slows down in the change of output. (C) shows the squared error quickly dropping to near zero, leaving only small components remaining. (E) is the final weight matrix of the 1st layer after learning, which correlates the error with the two rows of predictive ground sensors v_j. (F) is the weight matrix of the hidden layer and (G) of the output layer. (H) shows the Euclidean distance of the different layers from their starting point.

Overall, Fig. 4 shows that the network learns to use the predictive signals from the sensors in front of the robot to generate its steering output. This steering output slowly becomes stronger and then stabilises.

The squared average error in Fig. 4C slowly decays leaving only small spikes remaining. These spikes are mainly caused by the robot failing to learn one of the steeper bends (see coordinate 120×50 in D); this causes a spike in the error and overcompensation, causing learning in the other direction and so on.

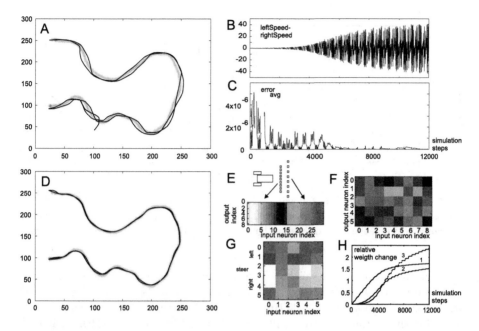

Fig. 4. Results of the line following task. (A) shows the robot at the very start of the simulation run from time step 0 to 600. (B) the difference between the robot wheel speeds just for the learned actions (i.e. the output of the DFL network): $v_l - v_r$. (C) the error signal squared: $error_{sq}$. (D) simulation run just before the end at 13400 to 14000 time steps. E-G show the weights of the different layers. The input neurons are on the x-axis and the output neurons are on the y-axis. (E) the weights of the 1st layer, (F) the weights of the 2nd layer and (G) the weights of the output layer. (H) shows the euclidean distance of the weights for each layer from their initial starting point.

The weights of the various layers are shown in Fig. 4E-G. The weights in the input layer (Fig. 4E) show a slow gradation from left to right as expected. Recall that there are two rows of predictive ground sensors in front of the robot; these cause two different weight maps which can clearly be seen. The inputs 0 to 14 correspond to the near ground sensor, and 15 to 31 the far sensor which looks further ahead, helping the robot predict bends better. These feed then into the hidden layer Fig. 4F and from there into the output layer Fig. 4G.

The overall weight development per layer over time is shown in Fig. 4H. The weights grow at a slower and slower pace but still continue because the robot is not able to completely avoid its reflex, and also probably because the system is non-linear. However, the error is very low even when the weights still change substantially; one could switch off learning if early stability is required.

We have gathered statistics of the line follower where we varied the learning rate over 4 orders of magnitudes, and evaluated how long it takes to stay below a certain error threshold for a specified time. For every learning rate two runs have been conducted with different random number seeds to test the dependence

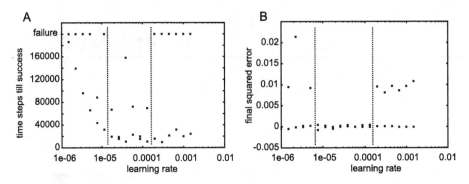

Fig. 5. Statistics of the line following task. (A) shows the relation between 'time to success' and learning rate. (B) shows the final squared error for different learning rates, and again between the dotted lines we have runs which resulted in low squared errors.

on the initialisation of the weights. The simulation was marked successful if the squared error Eq. 8 stayed below 10^{-6} for more than 2500 timesteps. The simulation was aborted after $200,000$ time steps if this criterion hasn't been reached. Figure 5A shows the time steps till criterion and Fig. 5B the resulting squared error. The time to reach the criterion decreases with higher learning rates, and is stable between learning rates of approx 10^{-5} and 10^{-4}, i.e. about one order of magnitude. Each simulation was run twice with different random seeds to test for initialisation effects. In the stable region this results in different times until success, whereas outwith this region about half of the runs fail. For lower learning rates the trial might still succeed eventually, e.g. Fig. 5B shows only three runs with low learning rates not converging. At low learning rates the robot sometimes learns essentially "open loop" with the error signal having very little impact, and learning can drift very slowly in the wrong direction only to be corrected later. At very high learning rates we get essentially one shot learning, which might accidentally learn the wrong behaviour. In the optimal regime (between the dotted lines) learning always arrives at low error values. Overall it's interesting that lower learning rates aren't necessarily better, but rather that intermediate learning rates perform best as they combine both, learning fast but slow enough so that the error signal can correct mistakes.

5   Shooter Game

In this scenario, we try to learn to play a first-person shooter purely from visual inputs. We use the Vizdoom (http://vizdoom.cs.put.edu.pl/) environment for this purpose, and train a controller to play against a single pretrained bot from Intel that ran in the Vizdoom 2016 competition. The setup was as follows:

The images returned from Vizdoom are RGB 160×120. We rendered the enemy in blue, and formed a reflex signal by finding the bounding box of the pixels closest to that colour. Note that the reflex is slow and also inherently noisy, as other events in the game are also rendered blue (e.g. the flashes that happen at respawns). The reflex also fails at times when the enemy is too distant or too close to the camera. For learning, we only supply the network with the greyscale image (Fig. 6G) flattened into a vector of 19200 inputs, so it is forced to discover purely spatial cues. The network has 2 fully-connected hidden layers of 5 units. The reflex is computed relative to the image centre, so a negative value implies the enemy is on the left. Shooting behaviour is entirely hardwired: if an enemy is detected within a threshold of the image centre, the bot fires. Note that the bot's only actions are to rotate in the plane, and shoot – it does not translate (but the enemy does). Instead of separate outputs for left and right, we have a single value produced by 3 neurons acting at different sensitivities:

$$\Delta\theta = g_{err}\,\text{error} + g_{net}\left(10v_0 + 3v_1 + v_2\right) \tag{9}$$

where $v_0, \dots, v_2$ are the network outputs, and $\Delta\theta$ is the change in orientation. g_{err} is $0.01g_{net}$, meaning that the maximum reflex ouput can only correct the error at a rate of 1% of the maximum network output: the reflex is slow.

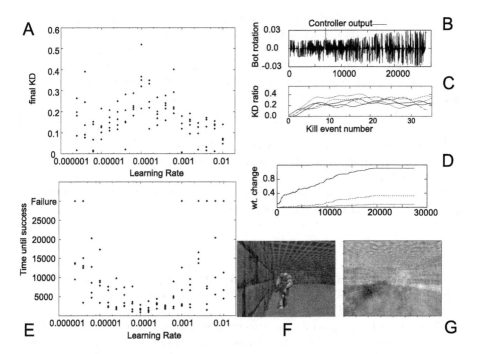

Fig. 6. (A) KD at end of trial. (B) $\Delta\theta$, the rate of change in bot's orientation against simulation step number, purely from the DFL output, in arbitrary units. (C) Example KD learning curves for 4 different runs - the time series of kills/deaths is filtered with a 2nd order lowpass filter to get a moving average, plotted over time. (D) Euclidean distance of weights for each layer from their initial point, against learning step. (E) Time to success vs learning rate - success is if the smoothed KD ratio reaches a threshold of 0.15. (F) Example input frame. (G) Example input weights.

5.1   Results

As before, we see that the controller outputs steadily increase over time (Fig. 6B). We initialise weights in a uniform distribution, and find that network produces significant outputs even at the beginning of each learning trial. With a high value of g_{net}, the bot can make very rapid aiming movements, reducing the error quickly and also generating exploratory movements even when the enemy is visually absent. On the other hand, it can cause the aiming to overshoot and oscillate around the target. The pattern of weights growth is less smooth than the Line Follower, due to the discontinuous nature of the error signal (Fig. 6D). The different scales are likely due to the large input size.

Unlike the Line Follower, it is not possible here to drive the error to zero, as the enemy will often abruptly appear somewhere in the image, causing an unpreventable error. One performance measure is simply how often our bot is killed vs how often it kills the enemy. In gaming, this is called the kill/death ratio, and we plot smoothed KD learning curves for 4 different runs in Fig. 6C. To perform statistical analysis, we measure the time taken to reach a threshold KD of 0.15 - Fig. 6E shows this as a function of learning rate. As before, there is a stable region in the middle where learning is consistently successful. For high learning rates learning is virtually instantaneous and approaches one shot learning. We also plot the smoothed KD for the final step of each trial (Fig. 6A); although a noisy measure, the same pattern emerges. Over time, the input weights blur the background, leaving dark and light blobs to detect the enemy and generate an aiming response (Fig. 6G). The learning rate in D, E was 0.0001.

Note that this is only one skill of a functioning FPS bot, as a real bot would require additional skills such as seeking rewards (e.g. finding health packs) and navigating the environment (e.g. avoiding collisions). The bot's restricted abilities lead to a relatively low KD ratio of 0.4, mainly caused by a high number of steps where it does not see the enemy. Future research will investigate whether the DFL approach can be used to acquire such skills, using the same fundamental approach of learning to anticipate a prewired behaviour.

6   Discussion

We have shown that a deep network which propagates its errors in a forward fashion from its inputs to its outputs is able to solve closed loop learning tasks. We have demonstrated this in both a first person shooter and a driving scenario.

Closed loop learning which aims to maximise a function or minimise one is usually referred to as reinforcement learning, where an agent learns to navigate an action space in a way that it maximises its accumulative reward [18]. The main drawback of this approach is the discrete state space which makes it hard to solve analogue problems. To overcome this problem, closed loop learning rules using correlation-based techniques [17] were introduced. These networks perform well in real robot tasks but are usually restricted to simple network architectures. DFL addresses this deficiency by introducing a deep architecture but staying firmly on correlation-based territory.

Plasticity has always been hotly debated in neurophysiology – the general understanding is that a large postsynaptic Calcium concentration causes LTP [1,6] and a low one causes LTD [8]. This requires a strong pre-synaptic drive to achieve a strong postsynaptic activity, and with that Ca influx [7]. In mathematical terms this would just lead to self-amplification of the synaptic weight, where strong presynaptic activity would lead to more postsynaptic activity and in turn stronger weights, and so on. However, suppose the learning signal and the actual activity were transmitted via the same synapse but were fundamentally separated [5], for example by using different frequencies: high frequency potentials could cause plasticity changes while low frequency potentials propagate behaviourally relevant activity [2]. DFL would provide here a mechanism which allows stable behaviourally relevant learning driven by heterosynaptic plasticity and Hebbian learning for the error signal which is also stable because it is constantly corrected by the error signal.

A different stance about synaptic plasticity (and ultimately how autonomous agents learn) has been taken by the deep learning community which recently claimed that error backpropagation is biologically realistic [4,13]. This has been demonstrated in a network with one hidden layer by introducing a separate feedback pathway from the output of the network to this hidden layer. While this shows promising results it still operates in an open loop fashion (i.e. output control), and thus contradicts the requirement for an agent to control its inputs.

Autonomous behaviour can only be understood by observing the whole loop [12], be it reinforcement learning [14] or correlation based learning [17]. Of course not all feedback loops traverse through the physical environment; they could be efference copies [3,16] or gating signals which are presented to minimise errors. However, even these signals need to be evaluated at the input of the agent, for example when stabilising an input image, the goal is itself an image, and therefore an input and not an output.

Our learning algorithm resembles similarities to work on inverse models in control generated by neural networks [15]. These models exhibit weaknesses when the mapping between direct and inverse control is not homogeneous. However, this weakness arises because backprop is trained by the error between the direct model and the inverse model because we need an *output*. Our model learns based on the error signal at the *input* after it has traversed through the environment which impacts in two ways: learning does not rely on a close relationship between inverse and direct model while at the same time it will develop far less rigid or autonomous solutions which are more compatible with a flexible agent than an engineering system. Plus the flexibility can be increased or decreased by adding or removing layers respectively.

Given that in DFL the error is propagated from the input to the output the novel aspect is that the more layers DFL has, the more it is remote from the immediate error feedback. This allows for flexibility which can be tuned by the number of layers, and demonstrates that the agent performs input control. The actions can be substantially varied as long as the error signal can be minimised and will be more pronounced with more hidden layers. In other words the more hidden layers we have the more behavioural flexibility the agent will show.

References

1. Bennett, M.: The concept of long term potentiation of transmission at synapses. Prog. Neurobiol. **60**, 109–137 (2000)
2. Canolty, R.T., Knight, R.T.: The functional role of cross-frequency coupling. Trends Cogn. Sci. **14**(11), 506–515 (2010). http://www.ncbi.nlm.nih.gov/pubmed/20932795
3. Grüsser, O.: Interaction of efferent and afferent signals in visual perception. A history of ideas and experimental paradigms. Acta Psychol. **63**, 3–21 (1986)
4. Lillicrap, T.P., Cownden, D., Tweed, D.B., Akerman, C.J.: Random synaptic feedback weights support error backpropagation for deep learning. Nat. Commun. **7**, 13276 (2016). http://www.ncbi.nlm.nih.gov/pubmed/27824044
5. Lindsay, G.W., Rigotti, M., Warden, M.R., Miller, E.K., Fusi, S.: Hebbian learning in a random network captures selectivity properties of the prefrontal cortex. J. Neurosci. Off. J, Soc. Neurosci. **37**(45), 11021–11036 (2017). http://www.ncbi.nlm.nih.gov/pubmed/28986463
6. Malenka, R.C., Nicoll, R.A.: Long-term potentiation – a decade of progress? Science **285**, 1870–1874 (1999)
7. Meunier, C.N.J., Chameau, P., Fossier, P.M.: Modulation of synaptic plasticity in the cortex needs to understand all the players. Front. Synaptic Neurosci. **9**, 2 (2017). http://www.ncbi.nlm.nih.gov/pubmed/28203201
8. Mulkey, R.M., Malenka, R.C.: Mechanisms underlying induction of homosynaptic long-term depression in area ca1 of the hippocampus. Neuron **9**(5), 967–975 (1992). http://www.ncbi.nlm.nih.gov/pubmed/1419003
9. Phillips, C.L.: Feedback Control Systems. Prentice-Hall International, London (2000)
10. Porr, B., von Ferber, C., Wörgötter, F.: ISO-learning approximates a solution to the inverse-controller problem in an unsupervised behavioural paradigm. Neural Comput. **15**, 865–884 (2003)
11. Porr, B., Wörgötter, F.: Isotropic sequence order learning. Neural Comput. **15**, 831–864 (2003)
12. Porr, B., Wörgötter, F.: What means embodiment for radical constructivists? Kybernetes, pp. 105–117 (2005)
13. Roelfsema, P.R., Holtmaat, A.: Control of synaptic plasticity in deep cortical networks. Nat. Rev. Neurosci. **19**(3), 166–180 (2018). http://www.ncbi.nlm.nih.gov/pubmed/29449713
14. Sutton, R.S., Barto, A.G.: Reinforcement Learning: An Introduction, 2nd edn. Bradford Books, MIT Press, Cambridge (1998)
15. Tejomurtula, S., Kak, S.: Inverse kinematics in robotics using neural networks. Inf. Sci. **116**, 147–164 (1999)
16. von Uexküll, B.J.J.: Theoretical Biology. Kegan Paul, Trubner (1926)
17. Verschure, P., Coolen, A.: Adaptive fields: distributed representations of classically conditioned associations. Network **2**, 189–206 (1991)
18. Watkins, C.J., Dayan, P.: Q-learning. Mach. Learn. **8**, 279–292 (1992)
19. Wörgötter, F., Porr, B.: Temporal sequence learning, prediction and control - a review of different models and their relation to biological mechanisms. Neural Comput. **17**, 245–319 (2005)

Deep Reinforcement Learning with Risk-Seeking Exploration

Nat Dilokthanakul[(✉)] and Murray Shanahan

Department of Computing, Imperial College London, London, UK
`n.dilokthanakul14@imperial.ac.uk`

Abstract. In most contemporary work in deep reinforcement learning (DRL), agents are trained in simulated environments. Not only are simulated environments fast and inexpensive, they are also 'safe'. By contrast, training in a real world environment (using robots, for example) is not only slow and costly, but actions can also result in irreversible damage, either to the environment or to the agent (robot) itself. In this paper, we consider taking advantage of the inherent safety in computer simulation by extending the Deep Q-Network (DQN) algorithm with an ability to measure and take risk. In essence, we propose a novel DRL algorithm that encourages risk-seeking behaviour to enhance information acquisition during training. We demonstrate the merit of the exploration heuristic by (i) arguing that our risk estimator implicitly contains both parametric uncertainty and inherent uncertainty of the environment which are propagated back through Temporal Difference error across many time steps and (ii) evaluating our method on three games in the Atari domain and showing that the technique works well on Montezuma's Revenge, a game that epitomises the challenge of sparse reward.

Keywords: Deep reinforcement learning · Risk-sensitive · Exploration

1 Introduction

Many exploration strategies employed in Deep Reinforcement Learning (DRL), such as ϵ-greedy, are highly inefficient due to their undirected nature [13,14]. Furthermore, statistically efficient methods that work well in low-dimensional settings [6,8,20] cannot be trivially extended to the domain of DRL. This inefficiency is one of the causes of DRL's high training cost.

In order to improve exploration in DRL, a variety of exploration heuristics have been proposed based mainly on the idea that exploratory actions should be directed towards observation of useful states or otherwise yield useful information. For example, Bellemare et al. [2] introduced a measurement called pseudocount, which assesses the novelty of the states visited. It is assumed that novel states contain more 'useful' information for learning. Alternatively, information gain can be measured with respect to changes in an agent's internal representation [12,18] and exploration can be directed towards maximising it. Similarly,

© Springer Nature Switzerland AG 2018
P. Manoonpong et al. (Eds.): SAB 2018, LNAI 10994, pp. 201–211, 2018.
https://doi.org/10.1007/978-3-319-97628-0_17

a heuristic called *optimism in the face of uncertainty* (OFU) encourages directed exploration towards states with high uncertainty estimate. Uncertainty can be estimated for different components of the model, for example, the environmental model [19] or the value function [16]. Intuitively, this heuristic aims to reduce the uncertainty of these quantities by treating them optimistically.

In this paper, we consider estimating an uncertainty of the return, i.e. the discounted sum of rewards. We propose a novel end-to-end algorithm which trains a deep neural network to represent the estimation of the return uncertainty. Next, we present arguments as to why applying the OFU heuristic with respect to this uncertainty estimate can be beneficial in many environments, and is especially well-suited to deterministic environments. Finally, we evaluate this method empirically in three of the most difficult Atari 2600 games, including the notorious Montezuma's Revenge and show that it outperforms the risk-neutral baseline.

2   Background

Let's consider the standard reinforcement learning (RL) setup [21]. Underlining our environment is a Markov decision process (MDP) which contains a set of states $s \in \mathcal{S}$, a set of actions $a \in \mathcal{A}$, a transition dynamic $P(s'|s,a)$, a reward function $R(s)$ and the discount parameter γ.

The goal of RL is to find a policy π that maximises the expected return. Let's define a return, G as a discounted sum of rewards which can be written as,

$$G = \sum_{t=0}^{\infty} \gamma^t R(s_t).$$

We can view G as a random variable which encapsulates the variability of simulated trajectories. Let's take an expectation of G with respected to the transition dynamics P and policy π given that we started at the state $s_0 = s$. We call this the *value* of the state s under policy π,

$$V^\pi(s) = \mathbb{E}_{\pi,P}[G|s_0 = s].$$

Similarly, we can define a state-action value or a Q-value as,

$$Q^\pi(s,a) = \mathbb{E}_{\pi,P}[G|s_0 = s, a_0 = a].$$

In RL, we are interested in achieving a policy π^* that maximises the expected return. One method is to find the optimal Q-value ($Q^* \equiv Q^{\pi^*}$) and then derive an optimal policy by choosing the actions with highest value ($a = \arg\max_b Q^*(s,b)$).

2.1   Deep Q-Networks (DQNs)

DQN [14] uses a deep convolutional neural network to represent an estimator $\hat{Q}$ which is optimised towards the optimal value Q^*. This estimator is trained by minimising the proposed DQN's loss which is a square of Temporal Difference (TD) error,

$$\mathcal{L} = (R_t + \gamma \max_b \hat{Q}_{\text{target}}(s_{t+1}, b) - \hat{Q}(s_t, a_t))^2,$$

with samples $\{s_t, a_t, R_t, s_{t+1}\} \sim D$ where D is a memory buffer containing trajectories collected from the simulation. The subscript 'target' denotes the use of a target network which is updated less frequently than the main network to stabilise the learning. This DQN's loss is derived from the Bellman's optimality equation,

$$Q^*(s, a) = R(s) + \gamma \mathbb{E}_P[\max_b Q^*(s', b)],$$

where we use s' for a state perceived after s. The correct value of Q^* has to satisfy this statement and, therefore, the loss aims to minimise the mismatch between the left and the right hand-side of the equation.

In this paper, we are interested in the variance of return $\text{Var}[G]$ which we refer to as risk. We want to train an estimator which estimates risk and use this estimate to help with exploration. Inspired by DQN, we look for a Bellman-like relationship for the risk value, so that we can derive a loss function for the neural network estimator that estimates the value of risk.

2.2   Variance of Return

Tamar et al. (2016) [22] derive a Bellman-like relationship for the second-order moment of return $M = \mathbb{E}[G^2]$. They then estimate the variance of G through the relationship,

$$\text{Var}[G] = \mathbb{E}[G^2] - \mathbb{E}[G]^2 = M(s) - V(s)^2.$$

The Bellman-like equation for M can be written as,

$$M^\pi(s) = R^2 + 2\gamma R\mathbb{E}[V^\pi(s)] + \gamma^2\mathbb{E}[M^\pi(s')]. \tag{1}$$

This statement is proven to hold for all $s \in \mathcal{S}$ and has a unique solution under an assumption that the policy is proper (see proof in [22]).

3   Deep Q-Networks with Variance Prediction

We extend Eq. 1 to incorporate a conditioning on $a_0 = a$ in the expectation. A simple modification to this statement can be written as,

$$M^\pi(s, a) = R^2 + 2\gamma R\mathbb{E}[Q^\pi(s, a)] + \gamma^2\mathbb{E}[M^\pi(s', \pi(s'))]. \tag{2}$$

This statement also has a unique solution which can be proven with a similar argument to that made by Tamar et al. [22]. The argument follows from the observation that $M(s, a)$ can be seen as the Q-value function of an MDP with reward $R^2 + 2\gamma R\mathbb{E}[Q^\pi(s, a)]$ and, therefore, the uniqueness of the Q-value function can be applied [4].

With the Bellman-like equation for $M^\pi(s, a)$, we are now ready to build a loss function for our neural network estimator $\hat{M}(s, a)$ which estimates the value of $M^\pi(s, a)$ where π is the current policy. We can then use $\hat{M}$ to estimate variance of the return $\mathrm{Var}[G]$ through the relationship $\mathrm{Var}[G|s, a] = M^\pi(s, a) - Q^\pi(s, a)^2$.

We follow the DQN algorithm to estimate $\hat{Q}$ which can also be viewed as an estimator of Q^π where π is the current policy that selects an action with the highest value, $a = \arg\max_b \hat{Q}(s, b)$. In addition to Q-value and M-value, we define a risk-seeking value, U, as a linear combination between the expected return and return variance;

$$U^\pi(s, a) = \mathbb{E}[G] + c(\mathbb{E}[G^2] - \mathbb{E}[G]^2), \tag{3}$$

where c is a hyper-parameter specifying the level of risk.

This risk-seeking value can be seen as a utility for the agent under a certain risk profile where c specifies the level of risk. For $c > 0$, the agent is risk-seeking and values risky states more than the states with high certainty in return. The agent is risk-averse when $c < 0$ and is risk-neutral when $c = 0$.

As suggested by Eq. 3, variance of the return can be calculated with the first- and second-order moment of the return. Accordingly, we extend DQN with two extra output streams resulting in a version with three output heads: one for the usual Q-value estimator $\hat{Q}$, another head for predicting the second-order moment, $\hat{M}$, and the last head for predicting the risk-seeking value, $\hat{U}$. We illustrate the architecture in Fig. 1.

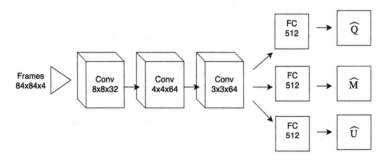

Fig. 1. The architecture of the network. Black arrows denote ReLU activation, boxes denote convolution layers and rectangles denote fully-connected layers.

In order to train each output head, we optimise the square loss of TD-error as done in DQN. The loss function for Q, M and U predictions are as follows:

$$\mathcal{L}_Q = (Y_Q - \hat{Q})^2 \tag{4}$$

$$\mathcal{L}_M = (Y_M - \hat{M})^2 \tag{5}$$

$$\mathcal{L}_U = (Y_U - \hat{U})^2 \tag{6}$$

where,

$$Y_Q = R_t + \gamma \hat{Q}_{\text{target}}(s', a')$$
$$Y_M = R_t^2 + 2\gamma R_t \hat{Q}_{\text{target}}(s', a') + \gamma^2 \hat{M}_{\text{target}}(s', a')$$
$$Y_U = R_t + \gamma \hat{Q}_{\text{target}}(s', a') + c\gamma^2 \max(0, \hat{M}_{\text{target}}(s', a') - \hat{Q}_{\text{target}}(s', a')^2),$$

s is the current state of the prediction, a is the action chosen and s' is the next state observed. a' is selected based on a greedy policy over the risk-seeking value,

$$a' = \arg\max_a \hat{U}_{\text{target}}(s', a).$$

The subscript *target* denotes that values are computed from the target network which gets updated less frequently than the prediction network in the same style as DQN. The gradient of each loss is also clipped at $[-1, 1]$ to avoid problems with outliers as suggested by Minh et al. [14].

Y_Q is the DQN target [14]. Y_M is defined by following Eq. 2 and Y_U is derived directly from Eq. 3 by expanding $G = R_t + \sum_{t'=t+1}^{\infty} \gamma^{t'} R_{t'}$. Finally, the $\max(0, .)$ operator is used to ensure the positivity of the variance. We can now carry out exploration under the guidance of the OFU heuristic by following a policy which selects actions that maximise $\hat{U}$ instead of the usual $\hat{Q}$.

4 Optimism in the Face of Risk

In this section, we explain why applying risk-seeking exploration, i.e. selecting actions with $\hat{U}$, is a justified, directed exploration strategy, especially in deterministic environments.

4.1 Optimism in the Face of Uncertainty

OFU is a well-known heuristic that is proven to be statistically efficient, for example in a multi-armed bandit setting [1]. The idea is simple; the more uncertain we are about the value of an action, the more it is important to explore that action to reduce this uncertainty. The trade-off between exploration and exploitation is done by, first assigning optimistic value to uncertain guesses and greedily following them. By collecting more data associated with the uncertainty, the guess becomes more accurate as the uncertainty is reduced.

4.2 Parametric Uncertainty and Inherent Uncertainty

As we aim to build a more accurate model, we want to reduce the uncertainty from imperfection in the model. This is called epistemic uncertainty or parametric uncertainty. Parametric uncertainty arises from imperfect information. Many models might be able to explain the data well but, with imperfect information, we are not certain as to which model is the correct one.

In MDP, there is also inherent uncertainty (aleatoric uncertainty) which arises from intrinsic randomness in the transition function and reward function. This randomness can give rise to risk, however it cannot be reduced. Therefore, risk-seeking exploration might not be suited for an environment with high inherent uncertainty.

4.3 Propagation of Parametric Uncertainty Through TD-Error

As a variant of Q-learning algorithm, we want to reduce parametric uncertainty of the estimator $\hat{Q}$. OFU can be applied to improve $\hat{Q}$ by encouraging exploration to the regions with high uncertainty of $\hat{Q}$. The main problem is how to estimate the uncertainty of $\hat{Q}$ in the first place, especially when a deep neural network is used to represent $\hat{Q}$. In order to avoid measuring uncertainty of $\hat{Q}$, we train an estimator that estimates the risk or the return uncertainty $\mathrm{Var}[G]$, and argue that this estimator readily subsumes the parametric uncertainty of $\hat{Q}$. It is important to stress that the uncertainty of $\hat{Q}$ is a different quantity to the uncertainty of G. Although, it is intimately related because $Q = \mathbb{E}[G]$.

Let's first discuss how can we directly measure the uncertainty of an estimator $\hat{Q}$. With recent advances in supervised learning with Bayesian neural network [5,17], it might seem trivial to put a prior on the neural network weights; as such, the uncertainty of the weights can then be learned automatically. However, with the use of TD-method, we train the neural network with next-step estimations as supervised targets. Therefore, techniques from Bayesian neural networks can only capture *local* uncertainty of the parametric error, while uncertainty from the target which uses a next-state estimate is not treated properly. A notable work by O'Donoghue et al. [16] proposed an uncertainty Bellman equation that propagates uncertainty of the Q-value estimator through multiple time steps. However, in their work, the uncertainty estimate is only represented with a linear function approximator with separately trained basis function. Moerland et al. [15] also suggested a way to propagate the parametric uncertainty through Bellman's update, but it was only evaluated on simple toy-tasks.

To see that the uncertainty of G contains the parametric uncertainty in the estimator $\hat{Q}$, we consider the following perspective made by [9,10],

$$G = \mathbb{E}[G] + G - \mathbb{E}[G] = Q + \Delta Q, \tag{7}$$

where

$$\Delta Q = G - \mathbb{E}[G].$$

ΔQ is a zero-mean residual which models the inherent randomness in the return distribution. This inherent uncertainty arises from the transition dynamics of the MDP and cannot be reduced. However, in a deterministic environment, ΔQ is zero and uncertainty in G can be attributed only by randomness from modelling error in Q. Since $\hat{Q}$ and $\hat{M}$ is learnt through TD-error, our estimator for $\mathrm{Var}[G]$ contains both parametric errors from $\hat{Q}$, $\hat{M}$ and the inherent uncertainty which propagates through multiple time-steps with TD-error.

Another perspective as to why modelling error can be measured from the risk estimator comes from the state aliasing effect [3]. The state aliasing effect occurs when the neural network perceives two distinct states to be the same, although they can lead to different outcomes. The aliasing states that lead to different outcomes can result in a high risk estimate. Therefore, it is a good idea to seek out these aliasing states to build a better representation, or learn to distinguish the two states. In contrast to count-based exploration [2] which tries to visit every possible state, risk-seeking exploration ignores aliasing states that lead to the same outcome, as it does not affect the uncertainty of the return. Therefore, exploration with modelling uncertainty is less wasteful than count-based exploration in principle.

However, exploration can only reduce the uncertainty until it cannot be reduced further. It is important to note that inherent uncertainty is not reducible and, thus, the agent will seek a sub-optimal behaviour if the inherent uncertainty is large. Therefore, our method works best in deterministic environments. Fortunately, this is not a restrictive assumption for many applications, since real-world classical physics is deterministic and many uncertainty comes from modelling error. This also leave an avenue for a future research where learning an estimator that can distinguish between inherent uncertainty and parametric uncertainty can be useful.

5   Experiments

5.1   Atari 2600

We evaluate our method in three Atari 2600 games, namely, Montezuma's Revenge, Seaquest and Venture. The games are considered to be high-dimensional with visual state as a $210 \times 160 \times 3$ array of RGB pixels. We pre-process the state by reshaping it into an 84×84 array and convert it into grey scale. We stack 4 consecutive frames together as a state input to the network. We also clip the reward to the range of $[-1, 1]$. The games are deterministic. However, we added no-op randomness [24] which introduces uncertainty to the starting state by performing a random number of no-op actions on initialisation.

Montezuma's Revenge is a notoriously hard exploration game in Atari 2600 domain where DQN and many of its variants fail to reach scores over 400 [11]. Only a few methods, built with novel exploration techniques, can cope with this extremely sparse reward environment [2,23]. The reason for this challenge is due to difficulties in reaching each reward, which needs a well-controlled sequence of actions to avoid the termination of the environment. We illustrate the difficulty of this environment in Fig. 2.

5.2   Effect of Risk-Seeking Exploration in Hard Exploration Games

In order to evaluate our method, we used ϵ-greedy with the greedy actions chosen according to the risk-seeking value, U. We observed the effect of risk on

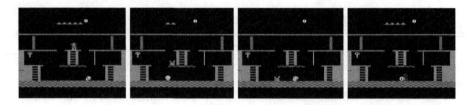

Fig. 2. In Montezuma's Revenge, there are many ways to terminate the game including falling down from a higher ground (middle figures) and getting killed by a monster (rightmost figure). In this room, the goal is to reach for the key which gives 100 scores and open a door with the key for an extra 300 scores. The agent needs to climb down stairs and jump over the monster to avoid losing a life.

three Atari games as we changed the hyper-parameter $c = \{0, 0.1, 0.2, 0.3\}$. We repeated each setting with three random seeds. At $c = 0$, the agent is risk-neutral and algorithmically similar to the normal DQN. As c increases, the agent assigns more value to states with larger return uncertainties and, thus, seeks more risk.

For each random seed, we saved agents at 100 million frames and evaluated them over 100 episodes with no-op evaluation [24]. We show the average results over three random seeds in Table 1. The σ sign denotes one standard deviation of the average scores.

Table 1. Average performances over three random seeds.

c	Montezuma's Revenge	Seaquest	Venture
0.0	1.3 $(\sigma = 2.3)$	5103 $(\sigma = 141)$	1057 $(\sigma = 162)$
0.1	262 $(\sigma = 227)$	4593 $(\sigma = 291)$	1199 $(\sigma = 106)$
0.2	133 $(\sigma = 230)$	4011 $(\sigma = 766)$	1194 $(\sigma = 162)$
0.3	556 $(\sigma = 963)$	4466 $(\sigma = 353)$	1174 $(\sigma = 103)$

The risk-seeking agents with $c = 0.3$ perform significantly better than the risk-neutral baseline in Montezuma's Revenge reaches the maximum score of 2500 with less than 100M frames. In contrast to the previously reported performances of DQN [11,14] on Montezuma's Revenge, none of vanilla DQN agent is able to reach the score as high as 500. This is suggestive that risk-seeking exploration has a positive benefit as it allows the agents to achieve high score that would not be possible with risk-neutral DQN. In this experiment, the risk-seeking factor c does not have observable effect on Venture and Seaquest.

Note that, the performances have high variation across random seeds as seen from high value of σ. Also, the experiments were repeated only with three random seeds. Therefore, the reported average scores are not statistically significant. We hypothesise that one of the factor contributinfg to the variation in performance is the random weight initialisation of neural network.

5.3 Reduce Sensitivity to Random Seed and Improve Performance with Biased Initialisation

In the previous experiment, we observed that the agent's performance varies greatly for different random seeds. We hypothesise that this is due to the random initialisation of the neural network which results in variations in the initial variance predictions of unknown inputs. At the start of the training, the variance is expected to be high due to the modelling errors, however, this is not the case for every initialisation. To alleviate this problem, we set the initialisation value of the bias of the fully-connected output layers of the second-order moment M and utility value U to 1.0. This results in average prediction of the variance at the start of the training to be approximately 1.0. We then train risk-seeking agent with $c = 0.3$ on Montezuma's Revenge, Seaquest and Venture with 10 random seeds.

Table 2. Average performance over 10 random seeds with initialisation variance bias set to 1.0

c	Montezuma's Revenge	Seaquest	Venture
0.0	0.8 $(\sigma = 1.6)$	4597 $(\sigma = 494)$	1125 $(\sigma = 147)$
0.3	1554 $(\sigma = 853)$	3900 $(\sigma = 633)$	1100 $(\sigma = 115)$

We found that this new initialisation scheme significantly improves the performance of Montezuma's Revenge. As shown in Table 2, the risk-seeking DQN with this initialisation has a much higher average score in this game. This result confirms that risk-seeking exploration helps in sparse-reward environment. However, for Seaquest and Venture, the performances of risk-seeking and risk-neutral agents converge to approximately the same scores.

6 Conclusion

In this paper, we consider the exploration-exploitation dilemma in DRL. In domains such as Montezuma's Revenge, methods with undirected exploration fail to learn successfully due to reward being sparse and delayed. With the aim to alleviate this problem, we propose a novel DRL algorithm based on DQN that can measure uncertainty of return or risk. This is achieved by learning the second-order moment at the same time as the expected value of return.

We motivate the use of risk estimate through an analysis that shows how the risk estimator contains parametric uncertainty through learning with TD-error. As a consequence, our method trades-off exploration and exploitation by encouraging exploration in areas that have high risk. In addition to being relatively simple, the risk estimator contains the modelling error, making it suitable for applying OFU. We also evaluate our method empirically and confirm that it is beneficial in a highly sparse reward environment – Montezuma's Revenge.

The short-coming of our method is that it works well only in deterministic environments and seeks sub-optimal solutions in stochastic environments. We leave the study of this method in stochastic environments for future work. An avenue for future research includes distinguishing inherent uncertainty from parametric uncertainty as we want to seek parametric uncertainty to reduce our estimation error while avoid the inherent uncertainty as it is irreducible. We highlight this as an interesting open problem.

Advances in distributional RL [3,7] estimates full distribution of the return which is much richer than the estimation of variance. It will be interesting to see how our method of risk-seeking exploration can be used to improve these methods further.

Acknowledgments. N.D. is supported by the Royal's Thai scholarships. We would like to thank Pin Aramwittaya, Pedro Mediano and Matthew Crosby for simulating discussions and comments on the paper. We would like to thank Nvidia for two Titan Z GPUs used in our experiments, and the reviewers for constructive feedback.

References

1. Auer, P.: Using confidence bounds for exploitation-exploration trade-offs. J. Mach. Learn. Res. **3**, 397–422 (2002)
2. Bellemare, M., Srinivasan, S., Ostrovski, G., Schaul, T., Saxton, D., Munos, R.: Unifying count-based exploration and intrinsic motivation. In: Advances in Neural Information Processing Systems, pp. 1471–1479 (2016)
3. Bellemare, M.G., Dabney, W., Munos, R.: A distributional perspective on reinforcement learning. In: International Conference on Machine Learning, pp. 449–458 (2017)
4. Bertsekas, D.P.: Dynamic Programming and Optimal Control. Athena Scientific, 2nd edn. (2000)
5. Blundell, C., Cornebise, J., Kavukcuoglu, K., Wierstra, D.: Weight uncertainty in neural network. In: International Conference on Machine Learning, pp. 1613–1622 (2015)
6. Brafman, R.I., Tennenholtz, M.: R-MAX-A general polynomial time algorithm for near-optimal reinforcement learning. J. Mach. Learn. Res. **3**, 213–231 (2002)
7. Dabney, W., Rowland, M., Bellemare, M.G., Munos, R.: Distributional reinforcement learning with quantile regression. arXiv preprint arXiv:1710.10044 (2017)
8. Deisenroth, M., Rasmussen, C.E.: Pilco: A model-based and data-efficient approach to policy search. In: Proceedings of the 28th International Conference on machine learning (ICML 2011), pp. 465–472 (2011)
9. Engel, Y., Mannor, S., Meir, R.: Bayes meets bellman: The Gaussian process approach to temporal difference learning. In: Proceedings of the 20th International Conference on Machine Learning (ICML 2003), pp. 154–161 (2003)
10. Ghavamzadeh, M., Mannor, S., Pineau, J., Tamar, A., et al.: Bayesian reinforcement learning: a survey. Found. Trends Mach. Learn. **8**(5-6), 359–483 (2015)
11. Hessel, M., Modayil, J., Van Hasselt, H., Schaul, T., Ostrovski, G., Dabney, W., Horgan, D., Piot, B., Azar, M., Silver, D.: Rainbow: combining improvements in deep reinforcement learning. arXiv preprint arXiv:1710.02298 (2017)

12. Houthooft, R., Chen, X., Duan, Y., Schulman, J., De Turck, F., Abbeel, P.: Vime: Variational information maximizing exploration. In: Advances in Neural Information Processing Systems, pp. 1109–1117 (2016)
13. Mnih, V., Badia, A.P., Mirza, M., Graves, A., Lillicrap, T., Harley, T., Silver, D., Kavukcuoglu, K.: Asynchronous methods for deep reinforcement learning. In: International Conference on Machine Learning, pp. 1928–1937 (2016)
14. Mnih, V., Kavukcuoglu, K., Silver, D., Rusu, A.A., Veness, J., Bellemare, M.G., Graves, A., Riedmiller, M., Fidjeland, A.K., Ostrovski, G., et al.: Human-level control through deep reinforcement learning. Nature **518**(7540), 529–533 (2015)
15. Moerland, T.M., Broekens, J., Jonker, C.M.: Efficient exploration with double uncertain value networks. In: Deep Reinforcement Learning Symposium @ Conference on Neural Information Processing Systems (2017)
16. O'Donoghue, B., Osband, I., Munos, R., Mnih, V.: The uncertainty bellman equation and exploration. arXiv preprint arXiv:1709.05380 (2017)
17. Pawlowski, N., Rajchl, M., Glocker, B.: Implicit weight uncertainty in neural networks. In: Bayesian Deep Learning Workshop at NIPS 2017 (2017)
18. Schmidhuber, J.: Curious model-building control systems. In: 1991 IEEE International Joint Conference on Neural Networks, pp. 1458–1463. IEEE (1991)
19. Stadie, B.C., Levine, S., Abbeel, P.: Incentivizing exploration in reinforcement learning with deep predictive models. arXiv preprint arXiv:1507.00814 (2015)
20. Strehl, A.L., Li, L., Wiewiora, E., Langford, J., Littman, M.L.: PAC model-free reinforcement learning. In: Proceedings of the 23rd International Conference on Machine Learning, pp. 881–888. ACM (2006)
21. Sutton, R., Barto, A.: Reinforcement Learning. MIT Press, Cambridge (1998)
22. Tamar, A., Di Castro, D., Mannor, S.: Learning the variance of the reward-to-go. J. Mach. Learn. Res. **17**(13), 1–36 (2016)
23. Tang, H., Houthooft, R., Foote, D., Stooke, A., Chen, O.X., Duan, Y., Schulman, J., DeTurck, F., Abbeel, P.: # exploration: a study of count-based exploration for deep reinforcement learning. In: Advances in Neural Information Processing Systems, pp. 2750–2759 (2017)
24. Van Hasselt, H., Guez, A., Silver, D.: Deep reinforcement learning with double q-learning (2016)

Online Gait Adaptation of a Hexapod Robot Using an Improved Artificial Hormone Mechanism

Potiwat Ngamkajornwiwat[1], Pitiwut Teerakittikul[1(✉)],
and Poramate Manoonpong[2,3(✉)]

[1] Institute of FIeld roBOtics,
King Mongkut's University of Technology Thonburi, Bangkok, Thailand
potiwat.n@gmail.com, pitiwut@fibo.kmutt.ac.th
[2] Institute of Bio-Inspired Structure and Surface Engineering,
Nanjing University of Aeronautics and Astronautics, Nanjing, China
poma@nuaa.edu.cn
[3] Embodied AI and Neurorobotics Lab, Central for BioRobotics,
The Maersk Mc-Kinney Moeller Institute, University of Southern Denmark,
Odense, Denmark
poma@mmmi.sdu.dk

Abstract. Walking animals show a high level of proficiency in locomotion performance. This inspires the development of legged robots to approach these living creatures in emulating their abilities to cope with uncertainty and to quickly react to changing environments in artificial systems. Central pattern generators (CPGs) and a hormone mechanism are promising methods that many researchers have applied to aid autonomous robots to perform effective adjustable locomotion. Based on these two mechanisms, we present here a bio-inspired walking robot which is controlled by a combination of multiple CPGs and an artificial hormone mechanism with multiple receptor stages to achieve online gait adaptation. The presented control technique aims to provide more dynamics for the artificial hormone mechanism with an inclusion of hormone-receptor binding effect. The testing scenarios on a simulated hexapod robot include walking performance efficiency and adaptability to unexpected damages. It is clearly seen that varying of hormone-receptor binding effect at each time step results in a better locomotion performance in terms of faster adaptation, more balanced locomotion, and self-organized gait generation. The result of our new control technique also supports online gait adaptability to deal with unexpected morphological changes.

Keywords: Artificial hormone mechanism · Gait adaptation
Autonomous robot · Adaptive behaviours · Gait generation

© Springer Nature Switzerland AG 2018
P. Manoonpong et al. (Eds.): SAB 2018, LNAI 10994, pp. 212–222, 2018.
https://doi.org/10.1007/978-3-319-97628-0_18

1   Introduction

Walking animals show various movement abilities including versatility, energy-efficiency, and adaptable locomotion to cope with uncertainty and to react quickly enough to changing environments. These inspire us to develop the many-legged robots to imitate natural properties of locomotion control of these walking creatures.

Different approaches, including machine learning algorithms, traditional engineering control techniques, have been utilized, aiming to achieve such a sophisticated locomotion. Furthermore, when there are unexpected situations such as hardware failures, it is crucial for the autonomous robots to recover their operations quickly and complete their task [1]. Currently, it remains a challenge to develop legged-robots that have effective adaptability without a complex control system [2].

The use of central pattern generators (CPGs) has been widely applied to the development of legged robots [3]. A single CPG was applied in some works [2, 4, 5]. Although implementing a CPG on a legged robot can yield sophisticated gait patterns, such a single CPG controller still faces a challenge to deal with leg malfunctions. Instead of using a single CPG, Ren et al. [6] and Barikhan et al. [7] developed multiple CPGs to enhance traversability and flexibility of a legged robot. Another promising method is the use of artificial hormone mechanisms [8, 9]. In [10], an artificial hormone network was introduced in helping an autonomous wheel robot to perform effective locomotion on unknown terrains as well as when faults occur in robot sensors.

Timmis et al. [11] proposed a combination of an artificial neural network and an artificial endocrine system (AES) to generate behaviour in artificial organisms. Here, what we improve is the AES to achieve online gait adaptation for a hexapod robot. According to this, the robot will be controlled by multiple decoupled CPGs with receptors and a hormone system. The improvement of the AES is achieved by an inclusion of receptor binding effect and resulted in not only self-organized locomotion generation but also online adaptation to unexpected malfunctions or leg damages.

The remainder of this paper is structured as follows: Sect. 2 presents the overall control mechanism where an improved artificial hormone mechanism (iAHM) and a multiple CPG model are introduced, Sect. 3 elucidates the experiment scenarios and parameters setting of the proposed model on a simulated hexapod robot, Sect. 4 demonstrates the experimental setup and discusses the results obtained from the experiments and the last section concludes the work and presents the recommendations for future work.

2   Hormone Mechanism

Hormones are regulatory substances released from endocrine glands into the blood vessel. When binding to specific receptors of their target cells, they trigger numerous cellular processes. The hormone mechanism is illustrated in Fig. 1A.

2.1   Artificial Hormone Mechanism (AHM)

Timmis et al. [11] proposed a bio-inspired mechanism called an Artificial Endocrine System (AES) for locomotion control.

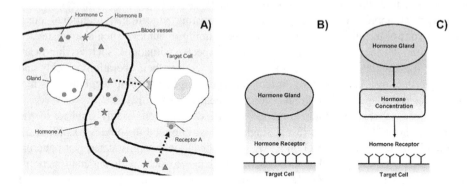

Fig. 1. (A) Hormone mechanism. (B) Standard artificial hormone mechanism. (C) Improved artificial hormone mechanism.

The main components of AES are hormone glands and hormone receptors as illustrated in Fig. 1B. In this model, hormone concentration at a particular time is calculated according to Eq. 1:

$$H_C(t) = \beta \cdot H_C(t-1) + \alpha \cdot H_G(t) \tag{1}$$

where $H_C(t)$ is hormone concentration at the current time step, $H_G(t)$ is a stimulation value at a current time step, α is stimulation rate and β is decay rate of hormone. From the Eq. 1, it can be seen that the changes of hormone concentration at each time step are dependent on the stimulation of hormones at current time step and the decay quantity of the hormone at the previous time step, $(t-1)$. The decay rate is equivalent to the metabolic clearance rate [10] which depicts the rate at which hormone is removed from the blood. The metabolic clearance rate is affected by many mechanisms [12], including metabolic destruction, binding of hormones to their receptors, excretion of hormones by the liver and excretion of hormones by the kidneys. In Timmis et al. study, these four mechanisms are presented conceptually as $\beta \cdot H_C(t-1)$ as shown in Eq. 1.

2.2   Improve Artificial Hormone System (IAHM)

As mentioned above, all four factors that influenced the metabolic clearance rate are presented as β in Timmis et al. [11]. In this paper, a new model is proposed, aiming to provide more dynamics for artificial hormones mechanism. The overview of the proposed system is shown in Fig. 1C, the metabolic clearance rate is expressed as the metabolic destruction and hormone-receptor binding effects indicated previously, while the excretion of hormones by the liver and kidneys are still not considered.

The hormone concentration at each time step of the improved mechanism is calculated as shown in Eq. 2:

$$H_C(t) = \beta \cdot H_C(t-1) + \alpha \cdot H_G(t) - H_R(t) \tag{2}$$

where $H_R(t)$ is a hormone-receptor binding effect. Additionally, the effect can be calculated as indicated in Eq. 3:

$$H_R(t) = \sum_{i=0}^{N} (\gamma_i \cdot Receptor_i(t)) \tag{3}$$

where γ_i is a binding effect rate for receptor i. $H_R(t)$ is the summation of a binding effect of each active receptor of cells influenced by the same hormone. However, $Receptor_i(t)$ is defined as a number of receptors in cell i at time t. There are two distinctive points where the improved artificial hormone mechanism can offer comparing to the system proposed in [10]. Firstly, the binding effects can influence hormone concentration. Secondly, each hormone receptor can be set to operate either at the same time or at the different time.

In order to examine the effect of the improved mechanism on a legged robot, a series of experiments is performed and illustrated in the next section.

3   Test Scenario

3.1   Mechanism for Implementation

In order to examine robot walking performance in the new proposed model, the simulation toolkit LPZROBOTS [13] based on the open dynamics engine was employed in a testing environment to represent the tested hexapod robot as shown in Fig. 2A. These six legs are identical, and each leg consists of three joints emulated from a basic

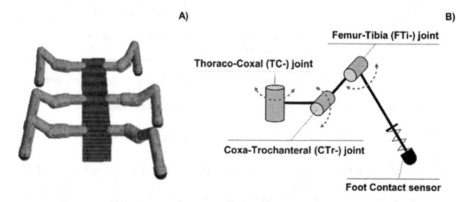

Fig. 2. (A) Hexapod robot in LPZROBOTS. (B) Example of the simulated robot components and motor joints location.

structure of a cockroach leg [14] without tarsus. Each robot leg consists of a thoraco-coxal joint (TC), a coxa-trochanteral joint (CTr), a femur-tibia joint (Fti), and a foot contact sensor (Fig. 2B).

3.2   Configuration and Parameters Setting

Both AHM and iAHM diagrams possess the same structure as illustrated in Fig. 3 except the iAHM has $H_R(t)$ in Eq. 2. For both AHM and iAHM, the hormone gland is stimulated from the correlation between six-foot contact signals (R1-R3 and L1-L3) and six motor commands (CR1-CR3 and CL1-CL3). After having been stimulated, the hormone is secreted and, subsequently, the hormone concentration rate will be changed. The hormone concentration will be increased when there is more orchestration of foot contact signals and motor commands.

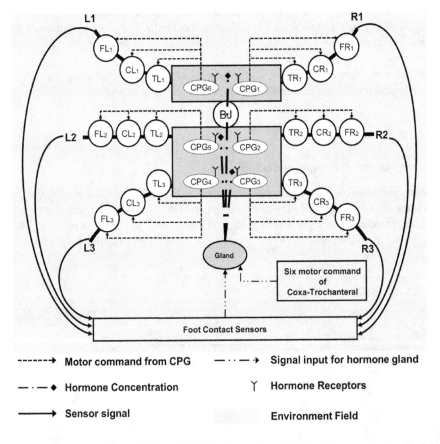

Fig. 3. The diagram of both AHM and iAHM with multiple CPGs of a hexapod robot. Each leg possesses a modulated CPG. The CPG has one receptor which binds the hormone and triggers the reactions of the target cell and subsequently change parameters for frequency control in the CPGs. Abbreviations are: TL, CL and FL = TC, CTr and FTi joints of the left leg while TR, CR and FR = TC, CTr and FTi joints of the right leg and 1, 2 and 3 = the front, middle and hind legs respectively and finally BJ = backbone joint.

In iAHM, when $H_R(t)$ is introduced into the Eq. 2, the hormone concentration can be varied according to $H_R(t)$ in a particular time step. Thus, the receptor on each leg can be set as either active or inactive at each time step, consequently, each leg can move adjustably in the online gait adaptation.

A CPG- based control system proposed by Manoonpong et al. [5] was adopted here to use in our experiment to generate basic rhythmic movement. The frequency control inputs that vary according to the hormone concentration operate the Module I through IV of the CPG (Fig. 4A). When the frequency control input increases, the angular velocity of the three joints will also increase. As a consequence, a CPG-controlled leg moves faster.

The relationship of the frequency control input and the hormone concentration is illustrated in Eq. 4:

$$Frequency\,control\,input = (0.17 * H_C(t)) + 0.02 \tag{4}$$

In our experiment, the frequency control input is set in a range of 0.02-0.19. The movement frequency of each leg is shown in Fig. 4B.

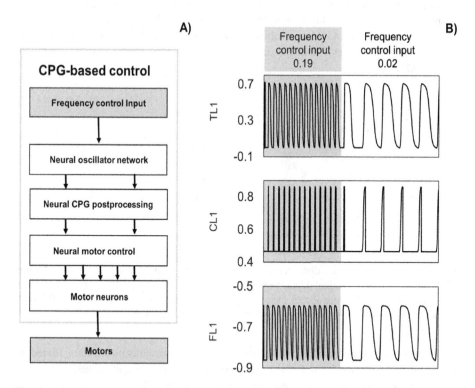

Fig. 4. (A) The diagram of a CPG-based control system on each leg of the robot. (B) The movement frequency of the joints on each leg when the frequency control inputs are 0.19 and 0.02. Notice that when the frequency control input is 0.19, each joint move faster than the frequency control inputs is 0.02, thus the leg also moves faster.

In all testing scenarios, the parameters for the stimulation rate, the decay rate, and the binding effect rate are 0.219, 0.2185, and 0.0005, respectively. Each receptor is set to be active at intervals of 20 steps starting from the receptor of L1, L2, L3, R1, R2 and R3 subsequently.

4   Experimental Setup and Results

4.1   Walking Performance and Self Organized Gait Generation

The aim of the first experiment was to indicate that the impact from varying of receptor stages could aid a robot in delivering better walking performance. In the first testing scenario, we compared iAHM with AHM by setting all initial parameters as indicated previously. We let the robot walk in forward direction for five minutes on the horizontal surface from the same starting point and orientation. Note that all legs on each side of the robot were initiated to move at the same phase. In iAHM, the receptors' status is set to be active as mentioned above while the receptors of AHM remained active in all time steps. We measured the distance that the robot moved along the X-axis and the swing magnitudes in Z-axis from both mechanisms. Each experiment was repeated for 10 times in order to investigate the variability of the performance.

It turned out that in iAHM, the robot was able to achieve almost three times greater of average walking distance and less standard deviation of the body swing compared to those of AHM (Fig. 5).

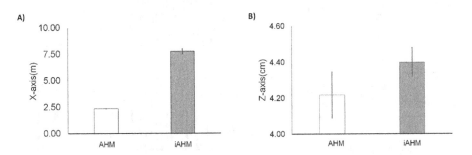

Fig. 5. (A) Average walking distances measured along the X-axis in AHM and iAHM were 2.34 m and 7.74 m and standard deviation were 0.01 m and 0.24 m, respectively. (B) Average of the center of mass height in the Z-axis measured when the robot walked in AHM and iAHM were 4.22 and 4.40 cm and standard deviation were 0.13 and 0.08 cm, respectively.

When the stance phase of the walking robot in both mechanisms was considered, it was found that all three legs of the same side were uplifted simultaneously in AHM since the legs of the same side possessed the same frequencies and phases (Fig. 6A). The robot walked jerkily and lost its body balance. The robot "heaved and rolled" simultaneously on the ground. The ground clearance of the robot during its movement varied wisely resulting in higher SD of the distance on the Z axis (Fig. 5B).

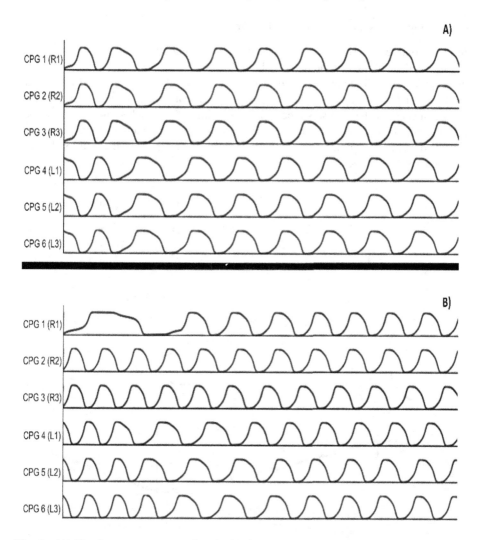

Fig. 6. (A) The frequency patterns of each side legs of the AHM robot (A) and iAHM robot respectively (R1, R2, R3 represents each right leg and L1, L2, L3 represents each left leg.)

In contrast with the robot in the iAHM, all legs of the robot were self-organized to automatically adjust to the variations of the activation stage of the receptors to obtain stable gait patterns. Then they paced in slightly different frequencies and phases according to their adjustment (Fig. 6B). The walking performance of the robot in iAHM is more adjustable than the robot in AHM due to the hormone-receptor binding effect (see Eq. 2).

4.2   Adaptability to Morphological Change

In the second testing scenario, we focused on the adaptability to an unknown damage of the tested robot. We let the robot walk with complete six-legs for 1.5 min. Then we disabled R2 and L3 leg as indicated in Fig. 7A and disabled R2, R3, and L3 legs as indicated in Fig. 7D and we allowed the robot to walk for 3.5 more minutes. The testing of each mechanism was repeated for five times. The result shows that the robot with the AHM could not adjust its walking performance and direction along the X-axis. It walked circularly and failed to move forward (black line in Fig. 7B and E for the two disabled legs and three disabled legs, respectively).

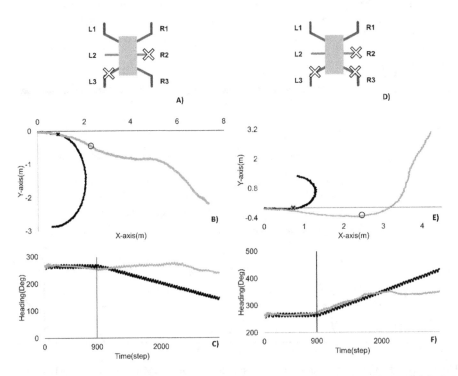

Fig. 7. Robot with disabled R2 and L3 legs (A) and robot with R2, R3 and L3 legs (D) failed to walk towards the desired trajectory (black line in B and E) and they also lost their orientation (black line in C and F) in AHM while they still kept their walking orientation in iAHM (gray line in B and E and C and F, respectively).

While the robot with the iAHM had an online learning capability to adapt its walking performance to walk towards the desired trajectory (gray line in Fig. 7B and E). When the movement orientation was considered, the robot with the AHM walked in a curved path (black line in Fig. 7C and F) while the robot with the iAHM tried to keep its orientation straight forward (gray line in Fig. 7C and F).

The hormone concentration of the robot in iAHM was investigated. Firstly, the normal hexapod robot was allowed to walk for 1.5 min, and the hormone concentration was calculated from the Eq. 2. The level of the hormone fluctuated highly for the first few seconds, then the hormone concentration fluctuated narrowly as shown in Fig. 8A. After being disabled for three legs, the hormone concentration was dropping significantly (see Fig. 8B). However, the robot spent a minimum time to adapt its gait to deal with the situation. When the robot achieved appropriate gait, the hormone concentration would increase again (see Fig. 8C).

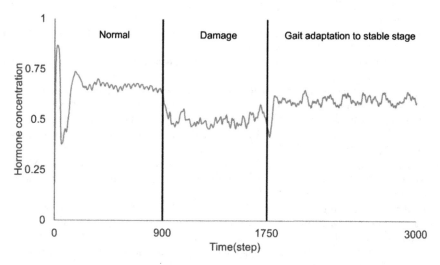

Fig. 8. The level of hormone concentration in each time step for the iAHM-driven robot

5 Conclusion and Future Work

We presented here a bio-inspired walking robot which was controlled by a combination of multiple CPGs and an artificial hormone mechanism with multiple receptor stages. The results are promising to prove that, the hexapod robot with iAHM can achieve a better locomotion not only faster but also more balanced locomotion compared to the previous AHM robot performance. The receptor activation in various times causes different frequencies and phases for gait generation. This enables the robot to move autonomously and adapt its gait pattern to handle morphological changes or damage. Although the walking performance of the legged robot has been improved by the stimulation of the hormone receptors at different phases, there are some limitations of the robot in achieving the walking performance of an insect.

Our future work will focus on conducting these experiments in real legged robots moving on different types of terrain to verify and confirm that an artificial hormone mechanism proposed in this study could improve walking efficiency and reduce the cost of transportation.

Acknowledgements. Poramate Manoonpong acknowledges funding by the Thousand Talents program of China and the research grant RGP0002/2017 from the Human Frontier Science Program (HFSP).

References

1. Chatzilygeroudis, K., Vassiliades, V., Mouret, J.-B.: Reset-free trial-and-error learning for robot damage recovery. Robot. Auton. Syst. **100**, 236–250 (2018)
2. Ijspeert, A.J.: Central pattern generators for locomotion control in animals and robots: a review. Neural Netw. **21**, 642–653 (2008)
3. Cully, A., Clune, J., Tarapore, D., Mouret, J.-B.: Robots that can adapt like animals. Nature **521**, 503–507 (2015)
4. Bernhard, K., Linnemann, R., Spenneberg, D., Kirchner, F.: Biomimetic walking robot SCORPION: control and modeling. Robot. Auton. Syst. **41**(2–3), 69–76 (2002)
5. Manoonpong, P., Parlitz, U., Wörgötter, F.: Neural control and adaptive neural forward models for insect-like, energy-efficient, and adaptable locomotion of walking machines. Front. Neural Circuits **7**, 12 (2013)
6. Ren, G., Chen, W., Dasgupta, S., Kolodziejski, C., Wörgötter, F., Manoonpong, P.: Multiple chaotic central pattern generators with learning for legged locomotion and malfunction compensation. Inf. Sci. **294**, 666–682 (2015)
7. Barikhan, S.S., Wörgötter, F., Manoonpong, P.: Multiple decoupled CPGs with local sensory feedback for adaptive locomotion behaviors of bio-inspired walking robots. In: del Pobil, A.P., Chinellato, E., Martinez-Martin, E., Hallam, J., Cervera, E., Morales, A. (eds.) SAB 2014. LNCS (LNAI), vol. 8575, pp. 65–75. Springer, Cham (2014). https://doi.org/10.1007/978-3-319-08864-8_7
8. Dufty, A.: Hormones, developmental plasticity and adaptation. Trends Ecol. Evol. **17**, 190–196 (2002)
9. Wingfield, J.C.: Endocrine responses to unpredictable environmental events: stress or anti-stress hormones? Integr. Comp. Biol. **42**, 600–609 (2002)
10. Teerakittikul, P., Tempesti, G., Tyrrell, A.M.: Artificial hormone network for adaptive robot in a dynamic environment. In: 2012 NASA/ESA Conference on Adaptive Hardware and Systems (AHS) (2012)
11. Timmis, J., Neal, M., Thorniley, J.: An adaptive neuro-endocrine system for robotic systems. In: 2009 IEEE Workshop on Robotic Intelligence in Informationally Structured Space (2009)
12. Guyton, A.C., Hall, J.E.: Pocket Companion to Textbook of Medical Physiology. Saunders, Philadelphia (2001)
13. Der, R., Martius, G.: The LpzRobots simulator. In: The Playful Machine. Cognitive Systems Monographs, vol. 15. Springer, Heidelberg (2011). https://doi.org/10.1007/978-3-642-20253-7_16
14. Zill, S., Schmitz, J., Büschges, A.: Load sensing and control of posture and locomotion. Arthropod Struct. Dev. **33**, 273–286 (2004)

Homeostatic Neural Network for Adaptive Control: Examination and Comparison

Oleg Nikitin[✉] and Olga Lukyanova

Computing Center, Far Eastern Branch of Russian Academy of Sciences,
Khabarovsk, Russia
olegioner@gmail.com, ollukyan@gmail.com

Abstract. Functioning of the biologically inspired neural network with cellular homeostasis is studied in the paper. The network is applied to the task of the control of agent behavior in the stochastic multi-goal environment. Importance of different aspects of the approach is studied on the setups with partially disabled features of the model. It is shown that only full model, incorporating both cellular homeostasis and homeostatically dependent weight correction rule led to the emergence of adaptive behavior. The proposed model is also compared to the $Q(\lambda)$ reinforcement learning on the same task with multiple goals. Results, illustrating the comparison between $Q(\lambda)$ and homeostatic neural network, show that proposed approach outperforms conventional in terms of adaptivity, quality of control and convergence speed.

Keywords: Cellular homeostasis · Neuronal plasticity
Adaptive behavior · Reinforcement learning
Biologically inspired neural network

1 Introduction

The interconnections and dynamics of the functioning of nervous systems are studied using modeling at all levels of abstraction, and different approaches to analysis. Nevertheless, in neuroscience, the level of abstraction can be divided into three dominant approaches to modeling:

1. High level of abstraction: formal neuron models [1], models of spiking neurons,
2. An average level of abstraction: biophysically plausible models of membrane processes [2],
3. High level of detail: models with protein molecular interactions [3].

Highly abstract models allow to obtain systems of effective machine learning and to simulate large-scale brain processes [4]. Models of the middle level make it possible to clarify the dynamics of the activation of the nerve cell, while detailed kinetic models allow a better study of the cellular processes controlling

© Springer Nature Switzerland AG 2018
P. Manoonpong et al. (Eds.): SAB 2018, LNAI 10994, pp. 223–235, 2018.
https://doi.org/10.1007/978-3-319-97628-0_19

this dynamics. Nevertheless, none of the approaches described above makes it possible to obtain neurocontrollers sufficiently reproducing the adaptive properties of biological nervous systems. The algorithms obtained as a result of its work require the specification of exact learning criteria and their adaptive dynamics depends on the system designer.

Unicellulars are capable of different kinds of learning, despite the absence of the nervous system. This is due to the presence of molecular training mechanisms in the cells. Such mechanisms are also present in the cells of multicellular organisms and allow the immune and nerve cells to adapt to changing conditions at the individual cellular level [5]. At the same time, in the modeling of neural processes, the most common approach is to represent a neuron as a passive information conductor.

Biological neurons are forced to survive throughout the life of the body, and for this they have to maintain cell integrity and cellular homeostasis. Cell activity depends on the process of maintaining homeostasis and regardless of the state of the environment of the neuron, it normally maintains a stable firing rate, while acting adaptively. Such properties of neurons are not usually modeled in adaptive control systems based on neural networks.

The homeostatic properties of biological organisms and their nervous systems cause them to function in a complex environment, achieving many goals and satisfying various needs.

In this paper, we tried to reproduce certain properties of biological nerve cells in order to obtain an adaptive controller of agent behavior and to model a conditionally "complex" environment with several motivations to test the ability of the proposed approach to act in a complex environment.

The homeostatic properties of neurons with respect to agent control were examined in [6]. In the paper, a network of neurons with the firing rate homeostais was used to control an agent in a two-dimensional world and perform the task of phototaxis. This approach allowed the adaptive reconstruction of the behavior programs after the change in the sensory system of the agent.

We continue to study similar systems. In the previous work [7], we presented an abstract model of the functioning of the nerve cell, which includes maintaining cellular homeostasis and homeostatic conditioned synaptic plasticity. A network of the proposed neurons was used to control a simple agent in a stochastic environment with two opposite motivations.

It seems useful to distinguish the role of different levels of modeling homeostasis in the adaptive properties of the neurocontroller. Thus, it is necessary to conduct a statistical analysis of the simulation data, analyze the interconnections that arise during the operation of the system and evaluate the hidden regularities of the functioning of the neural network. To compare the model with conventional methods, experiments with a similar setup in an environment completely identical to that used in [7] should be performed using the concept of $Q(\lambda)$ reinforcement learning.

Thus, the purpose of this work is to evaluate the functioning of the bioinspired model of the neuron and the neuronal network possessing homeostatic properties, to find out how the homeostatically dependent correction of the synaptic weights influences the training of the neural network, to draw some biological analogies for the simulated regularities, and to compare the model with conventional methods.

2   A Simulation Environment

To study the effect of the cellular homeostasis and synaptic plasticity in the neuron model, a simulation environment was developed. To investigate the adaptive properties of the model, the authors used an agent approach. The basis for the agent was a simple water polyp (hydra). Hydra has a simple neural network, distributed evenly throughout the body, which allows to model very simple structures. The agent network is also plain, and, in fact, represents three neural bunches: responsible for moving left, right and for capturing prey. Each bunch contains three neurons each (Fig. 1).

The agent (Fig. 1A) functions in a one-dimensional world with variable illumination and the shrimps that arise in its path (the agent's feed base, Fig. 1C), which it can eat. The agent can only move along the line to the right or to the left. The size of the world is 800 units, for a step the agent overcomes 5 units.

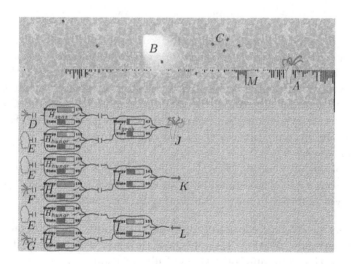

Fig. 1. Simulation environment. A - agent, B - directed light source, C - prays, D - touch sensor, E - stomach sensor, F - left light sensor, G - right light sensor, H - sensor neurons, H_{hungr} - hunger neuron, H_{sens} - touch neuron, I- actuator neurons, I_{grab} - pray catch neuron, J - effector of the catching, K - right mover, L - left mover, M - stripes, showing time agent was at the location.

The agent can eat the shrimps found on the way, thereby replenishing the energy reserve. In case of energy exhaustion, the agent dies and the simulation starts anew. The agent has a negative phototaxis - fear of light (the light source,

Fig. 1B). The network is constructed in a way to connect light sensors to the opposite side actuators. This leads to the negative phototaxis behavior of the agent. Thus, the agent has two opposite motivations: feeding and light avoidance. It allows to evaluate the adaptive behavior of the agent with two motivations.

The light source is directional and shines towards increasing the world's dimensional units (to the right). During the simulation, the light appeared in the center of the world (400 units) and changed its position to 3000, 6000 and 9000 steps for 300, 500 and 400 units, respectively.

The shrimps have a partially stochastic character of movement with positive phototaxis. In a greater light, they turn less and their runs per unit time become shorter. Thus, at the beginning of the simulation, the zone of the largest concentration of shrimps is from 400 to 800 units, and this zone narrows or expands afterward. This leads to a change in the location for a comfortable maneuver of the agent, since, the light dissipates at a distance of 150 units and it shapes the comfortable zone for the functioning of the agent. The agent can capture the shrimp only if the two conditions coincide: the shrimp will cross the agent's path, and the agent will send the command to capture the prey.

3   The Model of Homeostatic Neural Network

The agent is controlled by a simple neural network of three layers:

1. A layer of sensory cells, transmitting environmental indicators: light to the right, light to the left, touch of prey and fullness of the stomach (Fig. 1D,E,F,G). These cells generate a signal directly or inversely proportional to the monitored indicator. Thus, the stomach sensor generates a signal when the agent has not caught the shrimp for more than 50 steps.
2. A layer of sensory neurons, consisting of monopolar neurons (Fig. 1H), responsible for the primary processing of sensory information and functioning in accordance with [7].
3. A layer of output neurons (Fig. 1I), responsible for the integration of heterogeneous sensory information and activation of actions (movement to the left, to the right, capture of prey), functioning in accordance with [7].

The agent in all simulations initially was in the position of 450 units, in the zone of intensive illumination. Then, because the agent avoids the light, it moved to the zone with a minimum brightness of illumination.

The plasticity of the synapses is regulated by the rules from the model in [7]. Initially, neurons are not connected to the network and synapses are formed from stochastic co-activation of pre-synaptic and post-synaptic neurons. This process leads to a random order of bunches formation, responsible for moving to the right and to the left. So, if the bunch that transmits light signals from the right side to the left movement neuron is formed first - the agent will first move to the left, although the illumination is higher on the left. This will continue until both bunches are formed.

4   Options of the Experiment Setup

The stochastic process of synapse formation leads to the development of two options for the beginning of the simulation:

1. The agent moved to the left side of the world. On the left, the concentration of shrimps is extremely low, since the light is directed to the right. For the agent is difficult to go to the right, because light is located in the center and it is strongly repulsive. Thus, it is harder to survive for the agent.
2. The agent moved to the right side of the world. On the right, the concentration of shrimps is consistently high and it is easier to survive.

Such two options of the beginning of the simulation increase the space for studying the limits of model adaptability and allow to consider behavior of the agent not only in comfortable conditions, but also in extreme conditions.

Since the agent has a negative phototaxis, and the food base has positive phototaxis, for the survival of the agent it is sometimes necessary to exercise behavioral choice and to leave the space of light comfort in order to search of food. If the agent get on the right side of the world, it must pass through the zone of maximum illumination and get to the left.

All this allows to study the role of different components of the model in the adaptive behavior of the agent and make conclusions about the importance of cellular homeostasis and its role in synaptic plasticity for biological and model nervous systems.

To study the role of homeostatic properties of neurons in synaptic conductivity and adaptive ability of the neurocontroller, it seems useful to make simulation experiments with the model, gradually removing various levels of plasticity.

There are three simulation options:

1. Simulation of the agent controlled by the full network model from [7], with a homeostatically dependent correction of the synaptic weights of the neuron damaging input,
2. Simulation of the agent controlled by the network, without the phenomenon of homeostatically dependent correction of the damaging input, where Eqs. (14–16) from [7] is replaced by:

$$w_{i,j}(t) = w_{i,j}(t-1) + \Delta_{w_{i,j}}. \tag{1}$$

3. Simulation of an agent controlled by the network without plasticity, where all weights $w_{i,j}(t)$ are set to be 4.5, to move them above the threshold ($T_{spike} = 4$).

Below, we will refer to these three simulation options as simulations with the "full plasticity model", "without homeostatic correction" simulations and "no plasticity" simulations, respectively. The remaining parameters and dependencies in the three simulations were identical.

For each version of the simulations, 300 tests were conducted, 12,000 steps in each. A number of primary indicators reflecting the quantitative characteristics of agent activity, such as, for example, the average life duration of the agent, the average number of shrimps and caught shrimps in the environment, the average accumulated energy, and also the survival factor of the agent were evaluated.

During the simulations, the agent moved around the world, depending on the position of the light source and the location of the preys.

5   The Comparison of Plasticity Levels

In the case of an agent simulation with a full plasticity model, the agent moved almost uniformly throughout the space provided on the left side of the world and on the right, avoiding the light source. In the simulation without homeostatic correction, the agent predominantly remained in the right side of the environment, and in the absence of plasticity, it almost did not move to the left at all (Fig. 2).

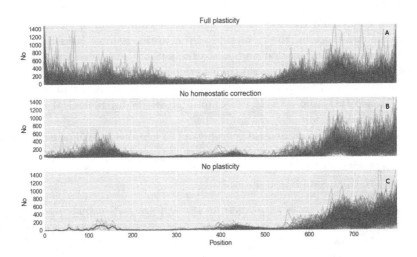

Fig. 2. The trajectories of the agent with the full plasticity model (A), without homeostatic correction (B) and without plasticity (C). The blue lines represent the results of individual simulations, the red line represents the mean. (Color figure online)

Due to the fact that, in the case without plasticity, synapses are created with weights above the threshold value, simultaneous co-activation of neurons often occurs and this leads to the early formation of synaptic connections. Because at the beginning of the experiment the intensity of light to the left of the agent is higher than to the right, in the case of simultaneous formation of synapses responsible for moving to the appropriate sides, the agent always moves towards the lowest light (in our case, to the right). Thus, in the case of fixed weights, the agent almost always chooses the right side, and the statistics from these simulations should be correctly compared not with all cases of formation of synapses (with full plasticity and without homeostatic correction), but only when the synapses responsible for moving to the right were formed first. That is why in the Fig. 3 only the right side cases are compared.

Simulation results for cases when in the network the bunch responsible for the right movement formed first are shown in Fig. 3. As can be seen from Fig. 3B1, in the case of a full plasticity model, the agent could cross a high-light region and move from the right side to left and back. This happened when all the shrimps on the side of the agent were eaten. In turn, in the case without a homeostatic correction and without plasticity, the agent could no longer change the location side. In the last two cases, the agent performed only motion-limited movements, and, on average, repeated the movements of the light source.

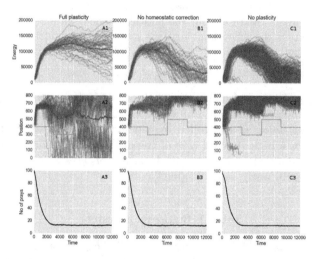

Fig. 3. Results of 300 simulations of adaptive behavior. Blue lines - results of particular simulations, red lines - average values, yellow lines - position of the light source. X-axes represent time steps. (Color figure online)

Figure 4 shows the dynamics of the average energy values of the agent in the process of 300 simulations. The graphs show that the agent with the full plasticity model effectively stored energy in all modeling options (blue line), while agents without plasticity and without homeostatic correction gradually lost energy, unable to efficiently accumulate it.

6 The Model Behavior in Different Sides of the Environment

The individual values of the indicators were calculated for each of the three simulations. The indicators are grouped according to which of the bunches of neurons responsible for the movement in one direction or another was formed

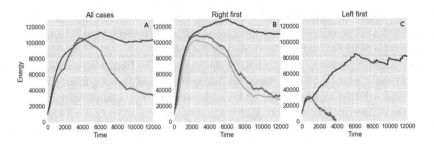

Fig. 4. Average agent energy values for 300 simulations. (A) All cases. (B) Sampling for the case when the bunch of neurons responsible for moving to the right was formed first. (C) Sampling for the case when the first formed a bunch of neurons, responsible for the movement to the left. The blue lines represent the results with the full plasticity model, the red lines represent the results for the plasticity model without a homeostatically dependent correction, the yellow line shows the results for the model without plasticity. (Color figure online)

first. So, it led to the initial movement of the agent in the appropriate direction, irrespective of the gradient of illumination.

Simulation of an agent controlled by a neural network with a full model of plasticity showed a high survival rate of the agent in the environment (survival rate for the entire sample was 0.89). Survival rates of cases with restricted control systems were significantly worse (Table 1).

Table 1. Statistical indicators

	Full plasticity			No homeostasis			No plasticity
	All	Left	Right	All	Left	Right	All
# of simulations	300	57	33	300	60	56	300
# when agent survived	267	57	22	50	32	0	132
Survival rate	0.89	1.0	0.65	0.32	0.53	0.0	0.44
Average life expectancy of the agent	11590	12000	10450	7058	10733	2451	10247
Average stored energy	92164	110297	52784	11105	17621	0	12297

The average lifetime of the agent with the full plasticity model remained stable, despite the primary choice of the right or left synaptic bunch. It was 11590 for all 300 simulations, 12000 for the case of a primary choice of the agent to go to the right and 10450 in the opposite case. At the same time, in simulations without homeostatic correction, the choice of the initial direction was critical for the life and functioning of the agent: the average life duration was 10733 for "right" cases and 2451 for "left".

During the experiment, the light source changed its position in space four times. Thus, it is possible to divide the entire time interval of the simulation into four phases corresponding to the four positions of the light source. The average values of the indicators were calculated from 300 simulation tests for each phase of the light.

Analysis of the performance of the agent in accordance with the phases of the position of the light source also confirms the conclusions made above. The agent with the full plasticity model adaptively functioned during all four phases, effectively accumulated energy and spent less energy on getting the food, while the agent without homeostatic correction and the agent without plasticity consumed a large amount of energy to search for food, but in the absence of internal homeostasis could not systematically replenish it and their average energy level was not only lower during the first two phases, but also significantly decreased in the third and fourth phases. It is important to note that in the first two phases, the agent with homeostatic correction collected energy faster than other agents, and in the third and fourth phases its energy did not practically decrease and kept at the level of 100 thousand units. This significantly exceeded its energy in the first phase.

There is no data for an agent without plasticity, in the case when the neuronal bunches responsible for moving to the left were formed first. This is due to the fact that by the time of the third phase, in absolutely all cases, the agent died, unable to adapt to the environment with a change in the light source.

Thus, the agent with the full plasticity model has much better adapted to the environment with different position of the light source than agents without homeostatic correction or without plasticity. This allowed it to save the energy received in later phases with less consumption.

Based on the foregoing, it can be concluded that the agent with the full plasticity model was able to adapt in the given environment and survive, regardless of its conditions, and its performance was superior to those of other simulations.

7 The Comparison with Q(λ) Learning Model

Models of learning with internal motivations are of current interest in the field of reinforcement learning, as they make it possible to obtain adaptive behavior of agent in a complex and stochastic environment [8].

To carry out a comparative analysis of the author's model with conventional methods, in an environment completely identical to that used in [7], a model with Q(λ) reinforcement learning was implemented.

To ensure the identity of the control tasks in Q(λ) reinforecement learning and the neural network, we chose Action and State values identical to the possible inputs and outputs of the neural network. Q(λ) reinforecement learning agent had the opportunity to choose three actions – move to the right, to the left and prepare to capture. The states of the agent were represented by indicators reflected by the activity of the touch sensors and the fullness of the stomach, the light from the left and from the right described in the Sect. 3. Q-values was

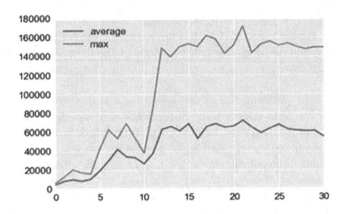

Fig. 5. Genetic optimization of Q(λ) reinforecement learning parameters. Y-axis represents energy values at the end of run, X-axis represents generations.

stored in the action-state table. For the negative phototaxis, the activation of light sensors was associated with a small penalty, which led to the choice of actions to avoid illuminated zones.

To ensure optimal adjustment of the hyperparameters, genetic optimization was used with a generation size of 40, with 30 generations, roulette selection and elitism with a size of 2. The parameters obtained were: $\epsilon = 0.08$, $\alpha = 0.3$, $\gamma = 0.55$, $\lambda = 0.5$. The optimization process is shown in Fig. 5.

300 simulations were conducted using both comparative control techniques. The development of situations in the simulated world took place in an identical manner and was determined by a change in the position of the light source in the environment (the yellow line in Fig. 6).

As can be seen in Fig. 6, the agent controlled by the "homeostatic" neural network was able to function steadily, despite changes in the environment, and also, combine the performance of two behavioral motivations - nutrition and avoidance of light. In Fig. 6A1, it can be seen that the agent, on average, steadily accumulated energy, both under conditions of a large number of shrimps, and in situations of their lack (Fig. 6A3). Figure 6A2 shows that the agent could move through a one-dimensional environment and approach the light source in search of food.

Meanwhile, the algorithm Q(λ) did not show the ability to adaptively reorganize behavior and was able to accumulate energy only under conditions of a rich environment (Fig. 6A1,A3).

The agent controlled by the Q(λ) model spent most of the time at maximum distance from the light source, unable to switch between motivations. This is seen in Fig. 6, and is particularly well illustrated by the time spent by the agent at certain points in the environment, which is shown in Fig. 7.

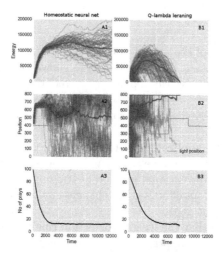

Fig. 6. Results of comparison of 300 simulations: A1-A3 - proposed model, B1-B3 - the model with $Q(\lambda)$ reinforcement learning. A1, B1 - values of energy accumulated by the agent in the process of simulations; A2, B2 - positions of the agent in the environment; A3, B3 - the number of shrimps. Blue lines are the results of individual simulations, red lines are mean values, yellow lines are the positions of the light source. The X-axis is time. (Color figure online)

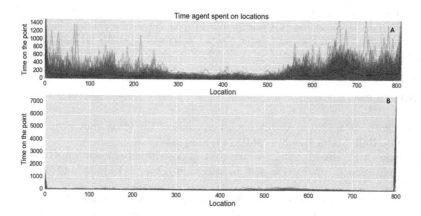

Fig. 7. Comparison of the frequencies of the agent at environmental points: A - the proposed model, B - the training model with $Q(\lambda)$ reinforcement learning (blue lines - the results of individual simulations, the red line - the average). (Color figure online)

Thus, the agent controlled by the proposed neural network could move around the one-dimensional world, and the agent controlled by the $Q(\lambda)$ model did not change positions, despite the dynamics of the environment (Fig. 7).

8   Conclusion

We analyzed the model of the neural network incorporating full homeostatically dependent weight correction. Statistical analysis showed that, with equal parameters, excluding the homeostatic dependent correction of the synaptic weights from the STDP plasticity rule leads to the inability of the agent to switch between mutually exclusive behavior programs and, as a consequence, ineffective functioning.

It is shown that the application of the author model of the neural network proposed in [7] allows to effectively manage the agent in a task with several motivations, while the conventional method of training with $Q(\lambda)$ reinforcement learning is not able to adapt effectively to changes environment and simultaneous achievement of two agent goals.

It can be concluded that the weight correction rule can be useful for application in the construction of adaptive systems, even on the basis of a simple neural network.

Perhaps, the use of such extended models of neurons supplemented by homeostasis will allow us to obtain more adaptive algorithms for machine learning and better understand the dynamics of self-organization processes in nervous systems.

Acknowledgments. The research was funded by RFBR project No. 18-31-00285. The calculations were carried out using methods and technologies which development was funded by RFBR according to the research project No. 18-29-03196. The computing resources of the Shared Facility Center "Data Center of FEB RAS" (Khabarovsk) were used to carry out calculations.

References

1. McCulloch, W.S., Pitts, W.: A logical calculus of the ideas immanent in nervous activity. Bull. Math. Biophys. **5**(4), 115–133 (1943)
2. Hodgkin, A.L., Huxley, A.F.: A quantitative description of membrane current and its application to conduction and excitation in nerve. J. Physiol. **117**(4), 500–544 (1952)
3. Kotaleski, J., Blackwell, K.: Modelling the molecular mechanisms of synaptic plasticity using systems biology approaches. Nat. Rev. Neurosci. **11**(4), 239–251 (2010)
4. Moren, J., Igarashi, J., Yoshimoto, J.: A full rat-scale model of the basal ganglia and thalamocortical network to reproduce Parkinsonian tremor. BMC Neurosci. **16**(1), 64 (2015)
5. Nilsonne, G., Appelgren, A., Axelsson, J., Fredrikson, M., Lekander, M.: Learning in a simple biological system: a pilot study of classical conditioning of human macrophages in vitro. BBF **7**, 47 (2011)
6. Di Paolo, E.A.: Homeostatic adaptation to inversion of the visual field and other sensorimotor disruptions. In: Meyer, J-A., Berthoz, A., Floreano, D., Roitblat H., Wilson, S.W. (eds.) From Animals to Animals. Proceedings of the Sixth International Conference on the Simulation of Adaptive Behavior, SAB 2000, pp. 440–449. MIT Press, Paris (2000)

7. Nikitin, O., Lukyanova, O.: Control of an agent in the multi-goal environment with homeostasis-based neural network. Procedia Comput. Sci. **123**, 321–327 (2018)
8. Singh, S., Barto, A.G., Chentanez, N.: Intrinsically motivated reinforcement learning. In: Proceedings of the 2004 Conference, NIPS 2004 Neural Information Processing Systems Foundation. Advances in Neural Information Processing Systems 17 (2005)

Collective and Social Behavior

Simulation of Heterogeneous Robot Swarm in Resource Transportation System

Seulgee Kim[✉] and DaeEun Kim[✉]

Biological Cybernetics Lab, School of Electrical and Electronic Engineering,
Yonsei University, 50 Yonsei-ro, Seodaemun-gu, Seoul 120-749, Korea
{slgee,daeeun}@yonsei.ac.kr
http://cog.yonsei.ac.kr/

Abstract. In this paper, we report simulation results of a heterogeneous swarm of robots using a potential field. The swarm models a resource transportation unit with supply units, defending units, and attacking units (enemy). Each class of vehicle unit possesses different maneuvering capabilities with different objectives. The supply units carry resources, the attacking units obstruct the supply units, and the defending units obstruct the attacking units. In this study, we verified whether our simulation works for the purpose (resource transportation system), and we also observed changes in the simulation results when the goal of each unit changed.

Keywords: Swarm · Heterogeneous swarm · Simulation
Resource transport system · Potential field

1 Introduction

Swarm robotics is a rapidly growing field in robotics. Swarm robotics is based on the concept that there are many advantages to using a group of relatively simple robots to accomplish a task over using a single highly complex robot [6–8]. In swarm robot control, there are two major fields into which swarms can be categorized: heterogeneous and homogeneous. Homogeneous swarm robot controls constitute robots that are completely identical, both physically and with respect to their control laws. To date, there has been a substantial amount of research on homogeneous swarms and their control methods [3,5,12,15,16]; however, there has been considerably less attention devoted to the field of heterogeneous swarms [4,10,13]. Heterogeneous swarm robot controls combine different types of robots that each have their own particular strengths. When these robots work together, the swarm displays the sum of the strengths of its individual units. Resource transportation systems need to be able to operate in areas far from the centralized command stations, where teleoperation becomes unfeasible because of communication constraints [9,12,16,17]. Furthermore, because of the hostile nature

© Springer Nature Switzerland AG 2018
P. Manoonpong et al. (Eds.): SAB 2018, LNAI 10994, pp. 239–249, 2018.
https://doi.org/10.1007/978-3-319-97628-0_20

of military operational environments, resource transportation system needs to be able to overcome the loss of individual members and, as a result, must not rely on a single "brain" robot to coordinate the swarm [1,3,5,14,15]. To increase the performance of the resource transportation system, the swarm will be heterogeneous, employing two types of robots: a supply unit and a defending unit. And there is an attacking unit that hinders transportation. The goal of this paper is to examine the validity of the presented models by applying the results of previous studies on successful homogeneous swarm techniques to heterogeneous environments and analyzing the results.

2 Method

This section introduces simulation environment and modeling for each class (supply units, defending units, attacking units), and describes how to verify the validity of the modeling.

2.1 Swarm Robot Modeling

In this paper, we propose a swarm modeling using a potential field. An artificial potential field based navigation is a reactive planning technique, where the immediate distances from obstacles are considered to compute the immediate move, without much concern for the future. In such a manner, immediate actions lead to motion of the robot, ultimately leading to the goal. All obstacles repel the robot with a magnitude inversely proportional to the distance, and the goal attracts the robot. The resultant potential, accounting for the attractive and repulsive components is measured and used to move the robot. The robot used in the simulation has five sensors as shown in Fig. 1, and the potential field uses the distance values of these five sensors. The first type of robot in the swarm is the supply unit. In resource transportation system, the supply unit's only job is to carry supplies from point A to point B; however, from a control standpoint, the supply unit is the backbone of the swarm. The second type of robot is the defending unit. The defending unit does not seek to drive to the endpoint but instead to maintain a defensive perimeter around the supply units. Lastly, an attacking unit that obstructs the supply unit was added to the simulation. The control laws for all of these units are based upon the requirements detailed in Table 1, and the variables used in the simulation are presented in Table 2.

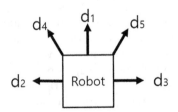

Fig. 1. Robot used in the simulation (it has five distance sensors).

Table 1. Unit control law design parameters

Supply unit	1. Move to goal (defined transportation goal)
	2. Avoid collisions with other units
	3. Avoid collisions with obstacles
	4. Change goal (from home to resource place, from resource place to home)
Defending unit	1. Move to goal (supply units' protection position)
	2. Avoid collisions with other units
	3. Avoid collisions with obstacles
	4. Obstructs the attacking unit approaching the supply unit
Attacking unit	1. Move to changing goal (supply units, obstruction position)
	2. Avoid collisions with obstacles
	3. Avoid collisions with other attacking units

Table 2. List of variables used in simulation

MS	Max robot speed
MST	Maximum speed change per unit time
MA	Max turn angle
SR	Robot size
R	Robot sensor range
MP	Minimum attractive potential at any point
SD	Robot safety distance
G	Goal position
S	Start position
AS	Scaling factor for attractive potential
RS	Scaling factor for repulsive potential
TD	A threshold distance, points within this threshold can be taken as same
DP	Degree of calculating potential
N	Number of robots

In the simulation in this study, the equation applied to each robot is as follows: first, the repulsive potential is given by Eq. 1, where c is the current direction, k is degree of calculating potential, and (d_1-d_5) are distance values obtained from 5 sensors. Moreover, the attractive potential is given by Eq. 2.

$$repulsive\ potential = (\frac{1}{d_1})^k \cdot [sin(c)\ cos(c)] + (\frac{1}{d_2})^k \cdot [sin(c-pi/2)\ cos(c-pi/2)]$$

$$+(\frac{1}{d_3})^k \cdot [sin(c+pi/2)\ cos(c+pi/2)] + (\frac{1}{d_4})^k \cdot [sin(c-pi/4)\ cos(c-pi/4)] \quad (1)$$

$$+(\frac{1}{d_5})^k \cdot [sin(c+pi/4)\ cos(c+pi/4)]$$

$$attractive\ potential = max([(\frac{1}{d_G})^k \cdot AS \quad MP]) \cdot [sin(ag)\ cos(ag)]) \quad (2)$$

where d_G is the distance from the goal, AS is the attractive potential scaling factor, MP is the minimum attractive potential, and ag is the angle of the goal. Finally, the total potential is given by Eq. 3.

$$Total\ potential = attractive\ potential - (repulsive\ potential \cdot RS) \quad (3)$$

where RS is the repulsive potential scaling factor. After calculating the total potential, the robot's next direction is obtained as follows.

$$preferred\ direction = atan2(current\ robot\ Speed \cdot sin(current\ direction)$$
$$+\ total\ potential(1), robot\ Speed \cdot cos(current\ direction)$$
$$+\ total\ potential(2)) - current\ direction$$
$$(4)$$

At this point, one must check whether the *preferred direction* obtained from Eq. 4 is an angle between π and $-\pi$. If $(preferred\ direction > \pi)$, subtract $(-2 \cdot \pi)$ from the *preferred direction*, and if $(preferred\ direction < -\pi)$, subtract $(2 \cdot \pi)$ from the *preferred direction*. Thereafter, checks if the *preferred direction* is greater than MA (max turn angle).

$$next\ direction = preferred\ direction + current\ direction \quad (5)$$

The next direction of the robot is obtained by adding the *preferred direction* and robot's current direction as shown in Eq. 5. The robot's next speed is obtained as follows.

$$next\ speed = sqrt(sum((current\ robot\ speed \cdot$$
$$[sin(next\ direction)\ cos(next\ direction)] + total\ potential)^2)) \quad (6)$$

Thereafter checks if the robot's *nextspeed* obtained from Eq. 6 is greater than MS (max robot speed) and MST (maximum speed change per unit time).

The above equations are summarized as follows: With knowledge of the current head direction and speed of the robot, it takes the distance value of the five sensors as inputs to determine the direction and speed of the next movement of the robot.

2.2   Simulation Environment

This simulation was implemented by MATLAB R2017b on a Intel(R) Core(TM) i7-5820K (16 GB RAM memory, 3.30 GHz CPU speed) with an Nvidia Geforce GTX 1080 Ti graphics card (11 GB video memory). The simulation environment is based on the resource transportation system. The simulation map is an 850 × 850 sized rectangle, where the resource ($x = 625, y = 625$) and home ($x = 175, y = 175$) are positioned diagonally to the corner of the square (Fig. 2(a)). Units are represented by a different color for each class (supply units: blue, defending units: green, attacking units: red) and the resources are displayed in orange when the supply unit carries resources (Fig. 2(b)). The robot was set to a square with a side length of 6. The size of the robot is set to be the same for all classes. The number of units used in the simulation in this study was 3 for each class (3 supply units, 3 defending units, 3 attacking units). It has different parameter settings for each class, as introduced in Sect. 2.1. The starting position of the robot is designated randomly near the home, and is generated with a certain distance for each class. The evaluation of the simulation results was used to evaluate the number of resources transferred by the supply units as a score, and all simulations were conducted on a 1000 time step basis. We first set each of the variables for our purposes to verify that our simulations work for the purpose.

Goal Position Variable Change Analysis Environment. A change in the G (goal position) variable was used to analyze the results of changing the position of defending units and attacking unit's goal of each class separately (See Table 3). In order to analyze the effect of the change of each G (goal position) variable on the simulation, first, only the goal variable of each unit is changed and the remaining variables are fixed to the value obtained in the simulation verification step. Additionally, repeated simulations of all types of units were carried out to check the results.

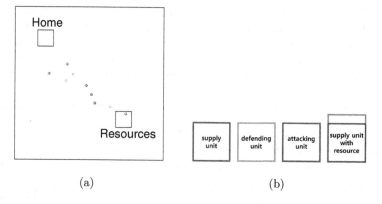

(a)   (b)

Fig. 2. (a) Simulation environment, (b) Unit shape for each class unit (Color figure online)

Table 3. Goal position variable change analysis case list

Case 1	When only supply units and attack units coexist/ Attacking units' goal: specified supply unit(man-to-man mark)
Case 2	When only supply units and attack units coexist/ Attacking units' goal: nearest supply unit
Case 3	When all supply units, defending units, and attacking units coexist/ Attacking units' goal: specified supply unit(man-to-man mark) and Defending units' goal: specified attacking unit(man-to-man mark)
Case 4	When all supply units, defending units, and attacking units coexist/ Attacking units' goal: nearest supply unit and Defending units' goal: specified attacking unit(man-to-man mark)
Case 5	When all supply units, defending units, and attacking units coexist/ Attacking units' goal: nearest supply unit and Defending units' goal: nearest attacking unit
Case 6	When all supply units, defending units, and attacking units coexist/ Attacking units' goal: specified supply unit(man-to-man mark) and Defending units' goal: nearest attacking unit
Case 7	When all supply units, defending units, and attacking units coexist/ Attacking units' goal: specified supply unit(man-to-man mark) and Defending units' goal: between nearest attacking unit and nearest supply unit
Case 8	When all supply units, defending units, and attacking units coexist/ Attacking units' goal: nearest supply unit and Defending units' goal: between nearest attacking unit and nearest supply unit

3   Results

3.1   Simulation Verification

Predicting the simulation results will naturally have the highest score when there are only supply units, and the lowest score when supply units and attacking units coexist. When both supply units, attacking units, and defending units coexist, scores will be higher than when supply units and attack units coexist, and scores will be lower than when only supply units are present. We have identified and adjusted the values of the variables so that our simulation can reflect the above predictions.

The following procedure was repeated to set the appropriate variable values.

Step 1. Add only the supply units to the simulation and adjust the value of the variable to maximize the score of the simulation.

Step 2. Add attacking units to the simulation and adjust the variable values to minimize the simulation score.

Step 3. Remove the attacking unit again from the simulation and check that the simulation is working properly (whether the score is somewhat similar to the highest score in the simulation).

Step 4. If it does not work properly in step 3, start from step 1 again. Conversely, if it works properly, go to step 5.

Step 5. Add defending units to the simulation and adjust the variables so that the points of the simulation increase more and more.

Step 6. Remove the defending unit again from the simulation and check that the simulation is working properly (whether the score is somewhat similar to the lowest score in the simulation).

Step 7. If step 6 does not work properly, start from step 5 again. Conversely, if it works properly, go to step 3.

The above procedure can be repeated several times to obtain appropriate variable values, and the values are shown in Table 4. Figure 3(a) shows the start of the simulation. As mentioned above, units of each class are randomly generated near the home at a certain distance. In Fig. 3(b), the defending units (green) protect the supply units (blue) carrying resources (orange) from the attacking units (red). The simulation results using the above variables are shown in Fig. 4. As you can see from the results, the simulation score was the highest when there were only supply units (score: 21), and the lowest when only supply units and attacking units coexisted (score: 9). Moreover when both attacking units, supply units, and defending units coexist, they scored lower than when there was only a supply unit, but received higher scores than supply units and attacking units coexisted (score: 15). This gave the same simulation results as our first

Table 4. Variables used in simulation verification and their value lists

MS	Supply units: 10/Attacking units: 20/Defending units: 20
MST	Supply units: 10/Attacking units: 20/Defending units: 20
MA	Supply units: 40/Attacking units: 40/Defending units: 40
SR	Supply units: 6/Attacking units: 6/Defending units: 6
R	Supply units: 300/Attacking units: 300/Defending units: 300
MP	Supply units: 2.5/Attacking units: 6/Defending units: 10
SD	Supply units: 6/Attacking units: 12/Defending units: 12
G	Supply units: from resources to home, from home to resources /Attacking units: nearest supply unit /Defending units: between nearest attacking unit and nearest supply unit
AS	Supply units: 300000/Attacking units: 300000 /Defending units: 300000
RS	Supply units: 1200000/Attacking units: 1200000 /Defending units: 1200000
TD	Supply units: 30/Attacking units: 30/Defending units: 10
DP	Supply units: 3/Attacking units: 3/Defending units: 3
N	Supply units: 3/Attacking units: 3/Defending units: 3

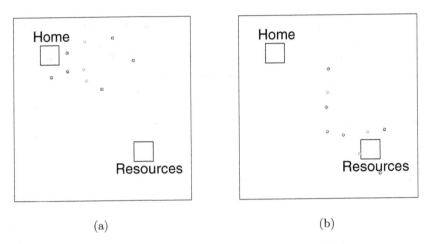

(a)   (b)

Fig. 3. Simulation scene. (a) Simulation start screen, (b) In the middle of a simulation, defending units (green) protects supply units (blue) carrying resources (orange) from attacking units (red). (Color figure online)

projections and shows that our proposed simulation method is suitable as a resource transportation system.

3.2   Results with Changes in G: Goal Position

In this section, we examine the change in score with goal position changes and analyze the reason for the score change. Figure 5 shows the change of the simulation score according to goal position change for each case in Table 3. When comparing the results of *Case* 1 and *Case* 2, when only supply units and attacking units coexist, when the attacking units' goal: specified supply unit (man-to-man mark) rather than when attacking units' goal: nearest supply unit, supply units are more obstructed and the simulation score is decreased. In all cases, when the attacking units' goal: specified supply unit (man-to-man mark), the simulation score is decreased to fit the purpose of the attacking units. In *Case* 3 and *Case* 7, when an attacking unit exclusively mark the supply units, it is better for the defending units to mark the attacking units exclusively.

In *Case* 4, *Case* 5 and *Case* 8, when an attacking unit does not mark the supply units as exclusive, then it has been confirmed that the defending units are obstructed by the exclusive marking (man-to-man mark) of the attacking units. Additionally, we analyze the effectiveness defending units based on the type of the attacking units. In *Case* 3, *Case* 6 and *Case* 7, the highest score in *Case* 3 indicates that when the attacking units exclusively mark the supply units, it is necessary for the defending units to mark the attacking units as exclusive, in order to increase the score (Fig. 5). Additionally, in *Case* 4, *Case* 5 and *Case* 8, it is necessary to increase the score to prevent the defending units from exclusively marking attacking units when the attacking units obstruct the nearest supply units, as the *Case* 4 score is the lowest (Fig. 5).

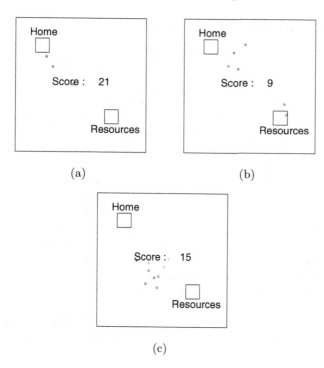

Fig. 4. Simulation results, Each simulation plot the score at the end of simulation. (a) Simulation results when there are only supply units, (b) Simulation results when supply units and attacking units coexist, (c) Simulation results when supply units, attacking units and defending units coexist (Color figure online)

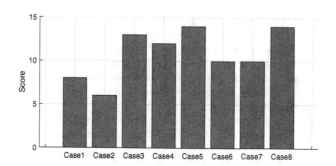

Fig. 5. Change of score according to change in G (goal position). Change of score according to G change in $Case$ 1 to $Case$ 8 (Table 3)

4   Conclusions and Future Work

In this paper, we succeeded in constructing a functional heterogeneous swarm that meets initial requirements using potential fields. Additionally, the simulation carried out in this paper shows that the use of heterogeneous swarms is both feasible and desirable for resource transportation system application and has the potential for a multitude of other applications. This study is different from existing potential field based formation control [2]. We use a potential field formed by using only five sensor values input to the robot. Rather than moving to the goal point while maintaining the formation, a heterogeneous swarm is constructed through the given goal position for each unit to show an effective transportation system process. Unlike conventional heterogeneous swarms, this study divides supply units and units that protect supply units from enemy units [11]. This is similar to the actual military transport system. This study shows that an efficient resource transportation system can be implemented even with simple rules in a simple robot system using only five sensors. We will continue to analyze the simulation results for observations of changes in the combination of variables and other variables that were excluded in this paper in a future study.

Acknowledgment. This work was supported by the National Research Foundation of Korea (NRF) grant funded by the Korea government (MSIT) (No. 2017R1A2B4011455).

References

1. Barnes, L., Alvis, W., Fields, M., Valavanis, K., Moreno, W.: Heterogeneous swarm formation control using bivariate normal functions to generate potential fields. In: IEEE Workshop on Distributed Intelligent Systems: Collective Intelligence and Its Applications, DIS 2006, pp. 85–94. IEEE (2006)
2. Barnes, L.E.: A potential field based formation control methodology for robot swarms. University of South Florida (2008)
3. Esposito, J.M., Dunbar, T.W.: Maintaining wireless connectivity constraints for swarms in the presence of obstacles. In: Proceedings of the IEEE International Conference on Robotics and Automation, ICRA 2006, pp. 946–951. IEEE (2006)
4. Iima, H., Kuroe, Y.: Swarm reinforcement learning method for a multi-robot formation problem. In: IEEE International Conference on Systems, Man, and Cybernetics, pp. 2298–2303. IEEE (2013)
5. Jadbabaie, A., Lin, J., Morse, A.S.: Coordination of groups of mobile autonomous agents using nearest neighbor rules. IEEE Trans. Autom. Control **48**(6), 988–1001 (2003)
6. Lee, W., Kim, D.E.: Adaptive division of labor in multi-robot system with minimum task switching. Simulation **8**, 10 (2014)
7. Lee, W., Kim, D.: Dynamic task allocation using a pheromone-based approach in factory domain applications. In: IEEE/WIC/ACM International Conference on Web Intelligence and Intelligent Agent Technology (WI-IAT), vol. 2, pp. 174–177. IEEE (2015)
8. Lee, W., Kim, D.E.: Local interaction of agents for division of labor in multi-agent systems. In: Tuci, E., Giagkos, A., Wilson, M., Hallam, J. (eds.) SAB 2016. LNCS (LNAI), vol. 9825, pp. 46–54. Springer, Cham (2016). https://doi.org/10.1007/978-3-319-43488-9_5

9. Lotspeich, J.T.: Distributed control of a swarm of autonomous unmanned aerial vehicles. Technical report, Air Force Inst of Tech Wright-Pattersonafb OH School of Engineering (2003)

10. Mabelis, A.A.: Wood ant wars the relationship between aggression and predation in the red wood ant (formica polyctena forst.) by. Neth. J. Zool. **29**(4), 451–620 (1978)

11. Moe, K.L.: There and back again; efficient foraging with heterogeneous agents using pheromones. Master's thesis, NTNU (2016)

12. Olfati-Saber, R.: Flocking for multi-agent dynamic systems: algorithms and theory. IEEE Trans. Autom. Control **51**(3), 401–420 (2006)

13. Pugliese, F., Acerbi, A., Marocco, D.: Emergence of leadership in a group of autonomous robots. PLoS ONE **10**(9), e0137234 (2015)

14. Scheutz, M., Schermerhorn, P., Bauer, P.: The utility of heterogeneous swarms of simple UAVs with limited sensory capacity in detection and tracking tasks. In: Proceedings of the Swarm Intelligence Symposium, SIS 2005, pp. 257–264. IEEE (2005)

15. Tan, Y.C., Bishop, B.E.: Combining classical and behavior-based control for swarms of cooperating vehicles. In: Proceedings of the IEEE International Conference on Robotics and Automation, ICRA 2005, pp. 2499–2504. IEEE (2005)

16. Tanner, H.G., Jadbabaie, A., Pappas, G.J.: Stable flocking of mobile agents Part I: dynamic topology. In: Proceedings of the 42nd IEEE Conference on Decision and Control, vol. 2, pp. 2016–2021. IEEE (2003)

17. Tanner, H.G., Jadbabaie, A., Pappas, G.J.: Stable flocking of mobile agents, Part I: fixed topology. In: Proceedings of the 42nd IEEE Conference on Decision and Control, vol. 2, pp. 2010–2015. IEEE (2003)

Bee-Inspired Self-Organizing Flexible Manufacturing System for Mass Personalization

Rotimi Ogunsakin[1]([✉]), Nikolay Mehandjiev[1], and César A. Marín[2]

[1] Alliance Manchester Business School,
Booth Street E, Manchester M13 9SS, UK
{rotimi.ogunsakin, n.mehandjiev}@manchester.ac.uk
[2] Information Catalyst for Enterprise Ltd., Crewe, UK

Abstract. One of the goals of Flexible Manufacturing System (FMS) is the mass production of personalized goods at cost comparable to the mass produced goods. This paradigm is referred to as mass personalization. To achieve this, the system has to seamlessly translate flexibility that can be achieved through the software that is responsible for the control of such system directly to the physical system, such that multiple distinct products can be produced in a non-batch mode. However, the present rigid design of Flexible Manufacturing Systems, which is characterized by static processing stations and rigid roll conveyor for part and material transportation, hampers this dream. In this paper, we propose a distributed architecture, which is implemented as Self-Organizing Flexible Manufacturing System (SoFMS), characterized by mobile processing stations that are capable of autonomously re-adjusting their location in real time on the shop floor to form an optimal layout depending on the mix of order inflow. This is achieved using the BEEPOST algorithm, an algorithm inspired by young honeybees' collective behavior of aggregation in a temperature gradient field. An agent-based simulation paradigm is used to evaluate the viability and performance of the proposed system. The result of the simulation shows that processing stations are able to autonomously and optimally adjust their location depending on the mix of order inflow using the BEEPOST algorithm. This capability also results in higher throughput when compare to a similar system with static processing stations. This approach is expected to engender the capability for production of one-lot-size order in FMS, which is a requirement for mass-personalization.

Keywords: BEEPOST algorithm · Flexible Manufacturing System
Mass personalization

1 Introduction

In mass personalization, Product family design consist of three types of modules: common modules that are shared across the platform; customized modules that allows customers to choose, mix and match; and personalized modules that allow customers to create and design [17, 18]. Whereas, Flexible Manufacturing System (FMS) is a

© Springer Nature Switzerland AG 2018
P. Manoonpong et al. (Eds.): SAB 2018, LNAI 10994, pp. 250–264, 2018.
https://doi.org/10.1007/978-3-319-97628-0_21

manufacturing system with the capability to respond to predicted and unpredicted changes in the production system, which may be due to variation in order mix or changes resulting from the internal mechanisms of the system's components [1]. In general, this flexibility is divided into two key categories. The first is resource (machine or processing station) flexibility, which is the ability of processing stations to make multiple parts, i.e. a single processing station can carry out multiple part production. Whereas, the second category, which is routing flexibility, is the ability to execute the same operation using multiple processing stations, meaning multiple processing stations can produce similar parts in the manufacturing system [2].

FMS are generally engineered for small lot-size production and with capability to adapt to limited variations in production designs and processes. Present FMS are yet to achieve the goal of being able to produce one-lot-size order in a non-batch mode, which is a requirement for mass personalization [3]. This may be as a result of the present FMS design paradigm, which is characterized by static processing resources (processing stations).

We explore relaxing this paradigm of static processing stations by introducing *mobile processing stations*, and instead of orders being transported by conveyor belt we introduced an autonomous product-agent capable of transporting parts to the mobile processing stations. We refer to this system as Self-Organizing Flexible Manufacturing System (SoFMS). The processing stations in SoFMS self-organize by adjusting their location, depending on the type of order-mix in the shop floor during manufacturing process. This enables the system to achieve seamless production process and higher throughput irrespective of changes in order mix distribution (external changes) or resource failure (internal changes). The self-organizing capability is achieved through the use of BEEPOST algorithm, an algorithm inspired by young honeybees' collective behavior of aggregation in a field with temperature gradient [4].

A simulation model is designed using multi-agent paradigm to evaluate the proposed self-organizing FMS. Analysis of the behavior of the processing stations during simulation (manufacturing) shows that the processing stations were able to self-organize by adjusting their location in synchronization with variation of order-mix inflow. This is achieved without any communication between the different processing stations on how to individually organize for optimal production, but by following simple rules that are inspired by young honeybees' collective behavior of aggregation in a temperature gradient field.

Furthermore, to evaluate the benefit of mobile processing stations and capability for self-organization through the use of BEEPOST algorithm, the performance of the proposed system is compared with that of a similar system with static processing stations. Result of the simulation shows that the proposed system records a higher production rate, lower lead-time per order and better reactivity to changes in order-mix during production process.

The capability of mobile processing station to adjust their location in synchronization with variation of order-mix inflow presents an alternative outlook into designing of FMS with the potential for production of one-lot-size order. This also have the potential to pave way for a new thinking in the design, layout and coordination of future manufacturing system to foster mass personalization, which is characterized by production of one-lot-size order.

The remaining of this paper is organized as follows. Section 2 reviewed related work in manufacturing systems and nature inspired algorithm for manufacturing system. Section 3 contains details of the proposed approach, which include architecture and algorithm. Section 4 contains the simulation and result of experimental evaluation of the proposed system. The final section contains the discussion and conclusion.

2   Related Work

Presently, there have been numerous approaches to engineer present manufacturing systems for mass personalization. A good example is the SmartFactort^{KL} for mass production of personalized soap with different color mixes [19] and production of personalized business card case [20]. The major success factor in SmartFactort^{KL} is modularization and "mechatronic changeability". This implies that the SmartFactort^{KL} is modularly structured and new production modules, which are mobile and portable, can be integrated and put into operation quickly to adapt to new production requirement. Bosch is already implementing this idea for size-of-1 manufacturing in their "Factory of The Future: Now, Next and Beyond" initiative. At the Hannover Messe 2018, Bosch presented what the company already offers (now) for connected factories, what solutions will soon be ready (next), and what it is developing for the future (beyond) [24]. The beyond is centered on mass personalization and the major initiative proposed by Bosch is that "the machines should also move" [24].

This is a relaxation of the core assumption of stationary production machines. However, the SmartFactort^{KL} approach is manual, but we investigate an approach whereby this process can be automated and individual manufacturing components are autonomous. What we also realize is that other approaches that maintained the core assumption have not recorded great success, i.e. those approaches where processing stations or machining stations are fixed in one position on the shop floor while the products move between these stationary-production machines [21–23].

However, flexible manufacturing paradigm is usually the approach of choice for product variety offering as required in mass personalization and has attracted considerable attention over the years. Flexible manufacturing system (FMS) is a manufacturing system with the capability to respond to predicted and unpredicted changes in the production system [1]. In FMS, flexibility is divided into two key categories, which are resource flexibility and routing flexibility. To achieve resource flexibility, individual processing station or resource has to be designed in such a way that they can adjust their internal mechanism to adapt to new changes in product designs and demand [10]. On the other hand, to achieve routing flexibility in a distributed scenario, both product and resources require interaction. Product needs to be aware of where the required resources are located, including alternative resources and the different possible routes to access these resources. This can be both physical and logical, which is referred to as physical and logical routing flexibility.

Physical routing flexibility is the actual layout of the possible route in the manufacturing system, such that multiple orders can be concurrently routed for multiple production process without physical obstruction. Whereas, logical routing requires a coordination mechanism for both optimal and alternative route selection, which can be implemented using central or distributed paradigm.

Logical routing flexibility is a major requirement to achieving mass personalization. This is because mass personalization scenario requires multiple products with different design, shape and color on the shop floor going through production process at the same time. Therefore requiring different resources and route at same or different time. Some of the available approaches used to achieve logical routing flexibility include: Market approach, where job agents (product agents) and resource agents act as buyers and sellers of resources in a virtual market respectively [5]. Here, cost and benefit calculation of the agent's activities is carried out for process and routing decision. Market/negotiation approach is further explored in control for Autonomous Agent for Rock Island Arsenal (AARIA) framework, where pricing is used as a negotiating protocol for allocating production capacity along a production chain [6]. The use of queue length, buffer space and process availability instead of price to allocate process to resources is also explored in [7] for achieving manufacturing flexibility.

2.1 Nature Inspired Approach

The use of a combination of computational and nature inspired approach has been applied to routing flexibility in FMS. For example, Petri-net based model for Internet of Things (IoT) enabled flow shop manufacturing with Ant colony optimization (ACO) as the coordination mechanism is explored in [8]. In this approach, products and resources are bond with RFID tags for resource and product visibility and location. Optimal process routes are determined by the use of Ant Colony Optimization (ACO). The use of stigmergy as a coordination mechanism for resource and process distribution is explored in [9] using the Product-Resource-Order-and-Staff-Architecture (PROSA) framework. It uses board as a generic mechanism to handle process changes. Information are dropped in the board by product and resource agents, these information placed on the board has a lifetime and refreshes at a specified interval to keep the information current for optimal decision-making. The use of Multi-Agent Systems (MAS) is explored for process coordination and resource allocation in [10]. The cooperative MAS was able to achieve effective resource allocation and coordination when each agent makes independent decision but adjusted by feedback of its own viability measure to maximize its own goal based on the limited information about the entire production system.

There also exists the use of solely nature inspired approach. For example, the concept of stigmergy has been applied to the design and coordination of manufacturing control system [11, 12]. A good example is the deposition of pheromones by foraging ants. In ant colony, pheromones are used as a coordinating mechanism by means of indirect or environmental mediated coordination [13]. This has been applied as coordination mechanism in multi-agent based manufacturing control system, and sometimes with PROSA as reference architecture for the development of prototypes [12].

Swarm of cognitive agents has also been applied to manufacturing system for controlling smart manufacturing system in order to overcome unexpected changes during production through autonomous reasoning and self-adaptation [14]. Autonomous Manufacturing System based on Swarm of Cognitive Agents (AMS-SCA) as proposed in [15], which shows real time adaptation to disturbances. In the AMS-SCA, the manufacturing system is considered as a swarm of cognitive agents where corresponding cognitive agents control work-pieces, processing stations, robots, and transporters. These approaches are still far away from enabling mass personalization capability in FMS. This is due to the inability of the production system to produce products of varied mix in a non-batch mode.

2.2   BEECLUST Algorithm

The BEECLUST algorithm is inspired by young honeybees' collective behavior of aggregation in a temperature gradient field [4]. A typical natural honeybee's hive is characterized by complex pattern of temperature fields, where the central brood-nest areas are kept at a comparably higher temperature (32 °C–38 °C) compare to the honeycomb areas and the entrance area, which have significantly lower temperature. The temperature in the brood-nest is most favorable for development of larvae and freshly emerged honeybee. Experimental observation shows that young honeybee (1 day old or younger) tend to aggregate preferentially in the spot most favorable for their development (at temperature between 32 °C and 38 °C) [16].

An experimental observation shows that the young honeybees were able to achieve a preferential aggregation into region where temperature is between 32 °C and 38 °C by following these simple rules [4]:

1. The warmer it is, the longer bees stay in a cluster
2. A bee that is not in a cluster moves randomly (maybe with a slight bias towards warmer areas)
3. If a bee meets another bee, it is likely to stop. If a bee encounters the arena wall, it turns away and seldom stops.

The BEECLUST algorithm is based on the above three rules. Different variations of the algorithm have been derived from these three simple rules with different levels of complexity and have been applied to swarm of Robots [4]. The BEEPOST Algorithm used in this paper is also inspired by the BEECLUST Algorithm.

3   Self-Organizing Flexible Manufacturing System (SoFMS)

The purpose of the research is to show how mobile processing stations are able to self-organize in a typical FMS with self-organizing properties using the BEEPOST algorithm. Manufacturing systems like the Self-Organizing Flexible Manufacturing System (SoFMS) with mass personalization capabilities are characterized by the potential for lot-size-of-one production, which implies that individual products will require distinct production plan, process and schedule. This also implies that the different products will require different resources at same or different time slot, different route leading to required resources, different operations to be performed on products at different time, and different logical operational sequence to arrive at the final product.

In the Proposed SoFMS, processing stations move during production depending on the mix of order in-flow, unlike in the conventional system where they remain stationary.

The processing stations represent aggregation of required resources for performing specific manufacturing process on products. The product-agents represents aggregation of mechanisms relating to product handling, transportation, routing, and mobility. This can be implemented as mobile robots or Intelligent automated guided vehicles (AGV) or Autonomous Intelligent Vehicles (AIV) with these capabilities (see Fig. 1).

The distributed nature of the problem space, which is the manufacturing system and the capability of the processing stations and products to move during production suggests the use of nature inspired approach in the coordination and control of the system.

To achieve this, the processing stations adjust their location depending on the variation of order-mix inflow. This implies that as the nature, designs and volumes of

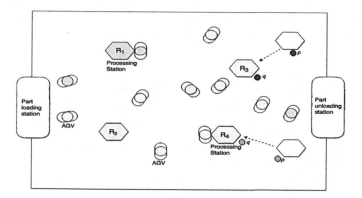

Fig. 1. Sample layout of Self-organizing Flexible Manufacturing System (SoFMS). During production, the resources or processing stations **R₃** and **R₄** re-adjust their respective location from **p** to **q** respectively using the BEEPOST algorithm

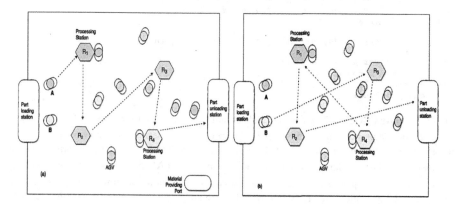

Fig. 2. (a) Shows a scenario of a process routing for product **A** and (b) shows scenario of a process routing for product **B**. It can be observed that the present configuration favors **A** with the route sequence $\{R_1, R_2, R_3, R_4\}$, while **B** with the route sequence $\{R_3, R_4, R_1, R_2\}$ will have to travel a longer route with higher traffic. Therefore, this position is optimal for **A** but not for **B**. Therefore, the processing station is expected to re-adjust its location using the BEEPOST in such a way that it will favor both **A** and **B**.

product to be manufactured changes, the processing stations re-organizes themselves and occupy a new location on the shop floor that is optimal for production of the new set of products (see Fig. 2). The processing stations achieve this by following the simple rules as expressed in the BEEPOST algorithm, and without any communication between the processing stations. This independent and autonomous self-organizing property gives robustness to the system, such that even during serious unexpected changes, the self-organization property of the system is unaffected.

3.1 BEEPOST Algorithm for Self-Organizing FMS

The BEEPOST algorithm implemented for the SoFMS is inspired by young honeybees' collective behavior of aggregation in a temperature gradient field. In the BEEPOST Algorithm, processing stations are named *resource-bees*, and instead of seeking for optimum temperature like in the honeybees' hive as implemented in the

BEEPOST Algorithm For Self-organizing FMS

```
1   procedure active_resource_bee(resource_bee_identifier)
2     k= resource_bee_identifier; p= product_identifier; s_0= get_initial_speed;
3     μ_0= set_initial_beespot; i= get_initial_movement_direction;
4       while (|s^p| ≠ Null && ω = false ) {ω → minimum overlaping distance}
5         foreach j ∈ N^k do read (N^p)
6           foreach j ∈ N_p^k do
7             Execute (production_plan);
8             Set e_t ← execution_time;
9             w_t ← Σw_t/n + e_t/Σw_t; {w_t → waiting time};
10            Wait (w_t);
11          end foreach
12        If (μ_0 = Null)
13            Set μ_0 ← N_p^k; {N_p^k → new beespot}
14            Set s_t ← s_0; {s_t → default speed (s_0)}
15        Else
16            Set s_t ← log[ρ((μ_0)^2 + (N_p^k)^2)^(1/2)]; {s_t → new speed } and ρ ∈ (0,1)
17            Set μ_0 ← N_p^k; {reset beespot}
18        end foreach
19      end while
20  end procedure
```

Fig. 3. The **BEEPOST** Algorithm for Self-organizing FMS. In **BEEPOST**, if there exist a product s^p in the production system (*step 1*). The *resource-bee* searches its neighborhood $j \in N^k$ for any product that it can execute. If there exist such product N^p in the neighborhood and it possesses a production plan executable by the *resource-bee* at N^k, then $N^k = N_p^k = \textbf{\textit{beespot}}(\mu_0)$ (*step 2–5*). Then the plan is executed by the *resource-bee* k, set the waiting time w_t, which is a function of the execution time e_t (*step 6–10*). If there are no previous *beespot* existing, the new beespot N_p^k is set as the first beespot μ_0 and speed is set as the default speed s_0. But if there exist a previous *beespot*, the new navigating speed s_t for the *resource-bee* is set as the logarithm of the product of the Euclidian distance between the present and the previous *beespot* and a speed bias ρ that controls how fast the speed increases depending on the size of the floor-space (*step 11–16*). Then the new *beespot* is now set as the initial beespot (*step 17*)

BEECLUST algorithm, they seek for *products*, which are the Intelligent AGVs with integrated mechanisms for product handling, transportation, routing, and mobility. The *resource-bees* wait at a spot referred to as *beespot,* a spot with higher concentration of products for which they are capable of executing a production process for during manufacturing process. This is achieved by resetting the *resource-bee's* waiting time at a *beespot* if the *resource-bee* executes any production process during the waiting period. Also, the *resource-bees* adjust their speed depending on the Euclidean distance between the present *beespot* and the previous *beespot* (See Figs. 3 and 4). The behavior of the *resource-bee* during manufacturing can be summarized as follows:

1. *Resource-bees* (processing stations) randomly move by setting their initial speed and direction.
2. If *resource-bee* meets a product with production process it can execute, it execute and set waiting time.
3. If waiting time is not over and a new production process is executed, *resource-bee* resets waiting time.
4. If waiting time is over and no production process is executed, step 1 is repeated.

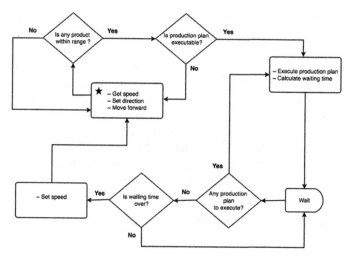

Fig. 4. State diagram for the BEEPOST algorithm (The * represent the start)

4   Experimental Evaluation

The system was implemented using agent based simulation software (Netlogo). Simulation experiment is designed to: (1) Operationalize and evaluate the viability of the proposed system and (2) to investigate the throughput gained by introducing mobility capability in the processing stations over a comparatively similar production system with stationary processing stations (stationary resources).

To investigate the above, two separate systems were developed and operationalized through the use of computer simulations:

1. The proposed Self-organized FMS (SoFMS) with mobile processing stations and **BEEPOST**, as the self-organizing mechanism
2. The base system (BS) with static processing stations.

A mass personalized shoe production scenario is used as a proof of concept. Shoe production is chosen because the production steps can easily be formulated and the facts that shoe manufacturing companies are currently investing in research on how to produce personalized shoes. For simplicity, we propose four processing stations for shoe parts production in the system with the following functions: (1) Make shoe-sole (2) Make shoe-head (3) Decorate shoe-head or Shoe-sole (4) Gum Shoe-head to Shoe-sole.

In each of the two simulations, maximum numbers of products, which are equivalent to number of products undergoing production in the production system, are set to 50, 100, 150, 200, 250 and 300. Each of these products is of 24 different product mixes, with maximum of 4 machining sequence per product. This implies there are 24 possible machining sequences (4-factorial). 30 simulation runs were performed in each of the developed simulation for 140,000 simulation steps each. The distribution of product mix in the system was skewed at interval of 20,000 simulation steps during the simulation to further observe the behavior of the system during unpredictable changes in product mix.

During simulation runs, the probability of a product-mix with a particular machining sequence being present in the system is set to the default in the first 20k simulation steps, which is $\frac{1}{24}$ (0.0417) – this is referred to as *mix-probability*. After 20k simulation steps, the *mix-probability* of one of the product-mixes is set to prob. = 0.5, while the rest 23 are set to prob. = 0.0217, which is the first scenario (scenario 1). At 60k–80k simulation steps, *mix-probability* of four of the product-mix are set to prob. = 0.2 and the rest 20 set to prob. = 0.01 (scenario 2). And lastly, *mix-probability* of seven of the product-mix were set to prob. = 0.125 and the rest 17 set to prob. = 0.0074 (scenario 3) at 100k–120k simulation steps. The different scenarios and the default scenarios are introduced interchangeably at an interval of 20k simulation steps. Table 1 shows the summary of the different scenarios.

Table 1. Probability distribution of product mix during simulations

Default		Scenario 1		Scenario 2		Scenario 3	
Mix probability	Mix type	Mix probability	Mix type	Mix probability	Mix type	Mix probability	Mix-type
0.047	1–24	0.5	1	0.2	1–4	0.125	1–7
–	–	0.0217	2–24	0.01	5–24	0.0074	8–24

The following parameters were measured in each of the two simulations setup to investigate viability and efficiency:

1. **Behavioral pattern:** This is the behavior of the processing stations (resource-bees) during the simulation. This is observed by visualizing the different positions occupied in the shop floor during production process using a heat-map plot.

2. **Production rate:** This is the average number of product produced per *10k* simulation steps during the simulation runs (*140,000* simulation steps in total).
3. **Average lead-time per unit:** This is the average time (measured in simulation steps) it takes to manufacture each product, i.e. the time each product spent in the production system.

4.1   Behavioral Pattern

After 30 different 140k simulation runs for the self-organizing FMS with mobile processing stations and BEEPOST as the self-organizing mechanism. The average coordinate of the different *beespots* occupied by the *resource-bee* (processing station) is visualized using a heat-map graph.

Figure 5(a) shows the movement pattern of the processing station when the initial speed was 1.0 step per simulation time and maximum order number in the shop floor is 50. The *resource-bees* were able to explore and self-organize by forming a consistent pattern on the shop floor. The processing stations moves close to the center region of the shop floor where there is higher concentration of orders to be processed more often than elsewhere on the shop floor. The 4 *resource-bees* autonomously aggregate at region that is optimal for higher production throughput by following the basic rules as expressed in the **BEEPOST** algorithm and without communication between the resources. Also, there exist non-overlapping constraints for the autonomous layout of the processing stations, which ensures they are at a minimum distance from each other.

Figure 5(b) shows the behavioral pattern of the *resource-bees* when speed was 1.0, with maximum order number in the shop floor set at 300. The *resource-bee* was also able to explore and autonomously move towards areas of higher product concentration. This shows that irrespective of the number of orders, the system is still able to self-organize and aggregate at location that is optimal for production processes.

4.2   Production Rate (PR) and Average Lead-Time Per Order (LT)

PR is observed to increase and LT decrease gradually at the beginning with the default scenario for both the Self-organized FMS (SoFMS) (*M50 to *M300) and the base system (BS) (F50 to F300) as shown in Figs. 6(a–f) and 7(a–f) respectively. However, The SoFMS recorded a relatively higher initial PR and lower LT compare to the BS. The introduction of scenario 1 at 20k–40k simulation steps (see Figs. 5 and 6) makes the PR to drop and LT to increase for both systems, but the impact on SoFMS is considerably lesser compare to that of the BS. The drop in PR and increase in LT is due the sudden change in product-mix distribution in the manufacturing system (scenario 1), a resemblance of what may occur in real-life due to sudden demand for some specific variant of designs. This has the most impact on the system because the system started practically producing one dominant order-type and requiring similar production plan, schedule and process. This lead to minimal flexibility in resource utilization and hence the resultant decrease in production rate.

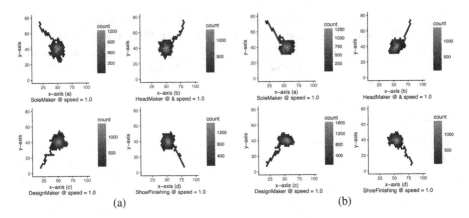

Fig. 5. Behavioral pattern of resources during 140k simulation time – The white region shows areas of less navigation activities compare to the heated regions (a) Processing station movement pattern at initial speed = 1.0 and order-number = 50 (b) Processing stations' movement pattern at initial speed = 1.0 and order-number = 300.

However, these sudden changes have less impact on the SoFMS compare to BS in all the setups. This is due to the ability of the processing stations in the SoFMS to self-organize by autonomously adjusting their location in real-time depending on the variation of order-mix using the **BEEPOST** algorithm and hence providing more flexibility to the system. The BS is incapable of this re-organization and hence the higher impact recorded.

When the default scenario was restored between 40k–60k simulation steps, PR is observed to increase and LT is observed to decrease for both systems, but SoFMS still recorded higher production rate. However, the introduction of scenario 2 and scenario 3 at 60k–80k and 100k–120k simulation time respectively has lesser impact on both systems. The impact on the SoFMS is still lesser compare to BS.

When numbers of orders are 50, 100 and 150 (see Figs. 6(a–c) and 7(a–c)), SoFMS records higher PR and lower LTO compare to BS throughout the simulation. This is due to the ability of the processing stations in SoFMS to aggregate at location that is optimum for production. This means that because there are fewer products in the system to be processed, the processing stations move close to these products by creating *bee-spots*, hence reducing the distance they have to travel by shortening the process route. Whereas, the processing stations in the BS are incapable of responding to these changes due to their immobility (Fig. 6).

When numbers of orders are 200, 250 and 300 (see Figs. 6(d–f) and 7(d–f)), from 60k to 140k simulation time, the PR and LT are approximately the same for both systems. This is due to increase in the number of products undergoing production process in the system. This means that number of product allocation per processing station will be higher; hence the processing stations in the SoFMS will have fewer tendencies to move i.e. to leave their *beespot*. This scenario makes the SoFMS behave almost like the BS and hence the similarity in their output.

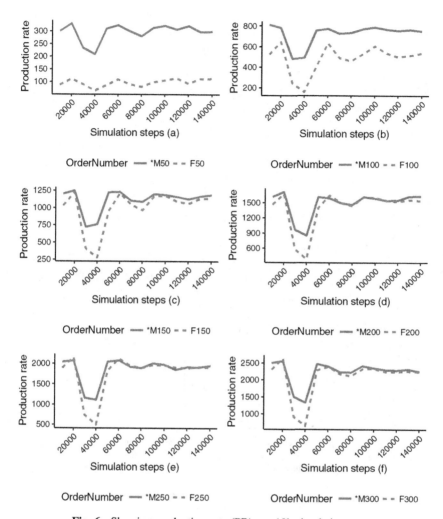

Fig. 6. Showing production rate (PR) per 10k simulation steps

We believe that the best scenario that depict real-life scenario is when the order number is between 50 and 150. The shop floor model used for the simulation has a limited dimension, so when there are more products, the movement of the processing stations is naturally constrained. We had to increase the number of products to 300 as a way of pushing the system to its limit for a clearer observation of the system's behavior and performance.

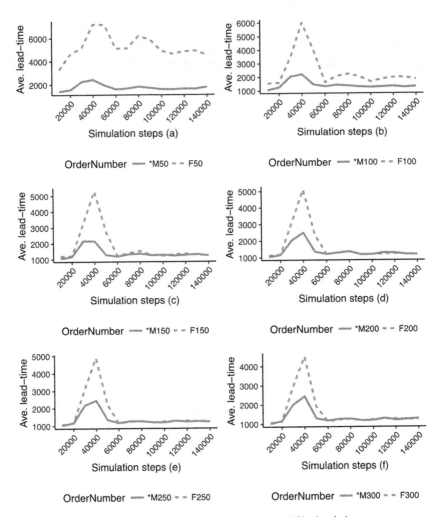

Fig. 7. Showing average lead-time per unit per 10k simulation steps

5   Discussion and Conclusion

Manufacturing systems flexibility is the major driving force for mass personalization. Different paradigms have sought to achieve the level of flexibility required to achieve personalized production in manufacturing environment. Example of such system is the Flexible Manufacturing System (FMS) and its variants where different approaches are used to achieve manufacturing flexibilities. The architectural and operational approach used limits the flexibility of these manufacturing systems, such that they are not entirely able to adapt to constant changes in product variation and volume. To address this gap we have proposed and implemented in silico, a decentralized manufacturing system with mobile processing stations called SoFMS.

Using computer simulation to operationalized SoFMS, is it observed that constant changes in product mix and volume generate cascades of events that disrupt the system's behavior during production, but the systems are observed to be able to keep-up production irrespective of product volume and constantly changing product mixes. However, the base system is able to achieve this level of adaptability as a result of decentralized architecture, where the products are capable of independent movement. But this flexibility does not suffice when the system experiences high variation in the distribution of product mixes. This is where the introduction of mobile processing station with self-organizing capability using the **BEEPOST** algorithm as implemented in SoFMS takes upper hand. The additional layer of flexibility introduced by the autonomous mobile processing stations in SoFMS gives the system better adaptablity to unexpected changes in the production environment and a considerably higher throughput compare to the base system.

Throughout the simulation, there was no observable instance where production rate equaled zero, even during unexpected changes in the production environment - like constant changes in product mix and volume. This shows that the SoFMS is capable of immediate adaptability to changes in product demands. These results suggest that the proposed concept is viable and put forward a new research direction in the design of production systems for mass personalization capabilities.

References

1. Scannon, P.J., Bernard, F., Tenerrowics, R.S., Dadson Jr., A.C.: Flex. Manuf. Syst. **5**(5), 917–921 (2011)
2. Shafer, S.M., Charnes, J.M.: Cellular versus functional layouts under a variety of shop operating conditions. Decis. Sci. **24**(3), 665–682 (1993)
3. El Maraghy, H.A.: Flexible and reconfigurable manufacturing systems paradigms. Flex. Serv. Manuf. J. **17**(4), 261–276 (2006)
4. Schmickl, T., Hamann, H.: BEECLUST: A swarm algorithm derived from honeybees. Bio-Inspired Comput. Netw. 95–137 (2010)
5. Lin, G.Y.J., Solberg, J.J.: Integrated shop floor control using autonomous agents. IIE Trans. **24**(3), 57–71 (1992)
6. Van Dyke Parunak, H., Baker, A.D., Clark, S.J.: The AARIA agent architecture: an example of requirements-driven agent-based system design. In: Proceedings of the First International Conference on Autonomous Agents, pp. 482–483 (1997)
7. Bussmann, S., Schild, K.: Self-organizing manufacturing control: an industrial application of agent technology. In: Proceedings of the Fourth International Conference on MultiAgent Systems, pp. 87–94 (2000)
8. Wang, M., et al.: A MPN-based scheduling model for IoT-enabled hybrid flow shop manufacturing. Adv. Eng. Inf. **30**(4), 728–736 (2016)
9. Van Brussel, H., et al.: Reference architecture for holonic manufacturing systems. Comput. Ind. **37**(3), 255–274 (1998)
10. Anussornnitisarn, P., Nof, S.Y., Etzion, O.: Decentralized control of cooperative and autonomous agents for solving the distributed resource allocation problem. Int. J. Prod. Econ. **98**, 114–128 (2005)
11. Heylighen, F.: Stigmergy as a universal coordination mechanism I: definition and components. Cogn. Syst. Res. **38**, 4–13 (2016)

12. Hadeli, et al.: Multi-agent coordination and control using stigmergy. Comput. Ind. **53**(1), 75–96 (2004)

13. Ricci, A., Omicini, A., Viroli, M., Gardelli, L., Oliva, E.: Cognitive stigmergy: towards a framework based on agents and artifacts. In: Weyns, D., Van Dyke Parunak, H., Michel, F. (eds.) E4MAS 2006. LNCS (LNAI), vol. 4389, pp. 124–140. Springer, Heidelberg (2007). https://doi.org/10.1007/978-3-540-71103-2_7

14. Park, H.S., Ur, R.R.Z., Tran, N.H.: A swarm of cognitive agents for controlling smart manufacturing systems. In: Proceedings of the International Conference on Natural Computation, pp. 861–867, January 2016

15. Park, H.S., Tran, N.H.: An autonomous manufacturing system based on swarm of cognitive agents. J. Manuf. Syst. **31**(3), 337–348 (2012). The Society of Manufacturing Engineers

16. Bodi, M., et al.: Interaction of robot swarms using the honeybee-inspired control algorithm BEECLUST. Math. Comput. Model. Dyn. Syst. **18**(1), 87–100 (2012)

17. Hu, S.J.: Evolving paradigms of manufacturing: from mass production to mass customization and personalization. Procedia CIRP **7**, 3–8 (2013)

18. Mourtzis, D., Doukas, M.: Design and planning of manufacturing networks for mass customisation and personalisation: challenges and outlook. Procedia CIRP **19**, 1–13 (2014)

19. Qin, J., Liu, Y., Grosvenor, R.: A categorical framework of manufacturing for Industry 4.0 and beyond. Procedia CIRP **52**, 173–178 (2016)

20. Gorecky, D., Weyer, S., Koster, M., Florek, S., Richter, D., Reboredo, R., Paral, T., Schigli, E., Esspenberger, T.: SmartFactoryKL System Architecture for Industrie 4.0 Production Plants (2016). http://smartfactory.de/wp-content/uploads/2017/08/SF_WhitePaper_1-1_EN.pdf. Accessed 11 June 2018

21. Onori, M., Semere, D., Lindberg, B.: Evolvable systems: an approach to self-X production. Int. J. Comput. Integr. Manuf. **24**(5), 506–516 (2011)

22. Onori, M., Lohse, N., Barata, J., Hanisch, C.: The IDEAS project: plug & produce at shop-floor level. Assem. Autom. **32**(2), 124–134 (2012)

23. Sanderson, D., Chaplin, J.C., De Silva, L., Holmes, P., Ratchev, S.: Smart manufacturing and reconfigurable technologies: towards an integrated environment for evolvable assembly systems. In: Proceedings of the IEEE 1st International Workshops on Foundations and Applications of Self-Systems, FAS-W 2016, pp. 263–264 (2016)

24. Bosch: #HM18 Factory of the Future Show - Bosch Rexroth [video online] (2018). https://www.youtube.com/watch?v=HNK7By5W0e8. Accessed 11 June 2018

Author Index

Printed in the United States
By Bookmasters